FUNDAMENTALS OF
GEOGRAPHIC INFORMATION SYSTEMS:
A COMPENDIUM

William J. Ripple, Editor

COVER IMAGE—Digital-elevation map of North America. The original data was collected as points during a gravity survey. The Geophysics Division of the Geological Survey of Canada interpolated the points into a raster format. The image was processed through a PC-based PCI image analysis system, and TYDAC's SPANS.

The minimum level shown is the dark blue (may appear as black) areas of the oceans, showing a depth of −3,264 m. From there, the depth grades upward through lighter shades of blue, reaching sea level, as represented by dark green. From sea level, the elevations continue upward through greens, yellows, oranges, browns, and reaches the maximum of +2,988 m in the red areas.

ISBN 0-944426-60-3

Printed in the United States of America

Published by
American Society for Photogrammetry and Remote Sensing
and
American Congress on Surveying and Mapping
5410 Grosvenor La.
Bethesda, MD 20814

FUNDAMENTALS OF GEOGRAPHIC INFORMATION SYSTEMS: A COMPENDIUM

TABLE OF CONTENTS

SECTION 5
GIS TRENDS

SECTION 6
GIS APPLICATIONS

SECTION 7
GIS SELECTION

SECTION 8
GIS LITERATURE AND NEWSLETTERS

PREFACE

We are faced with an increasing number of resource management and environmental problems. These include possible global climate change due to the greenhouse effect, acid deposition stemming from the burning of fossil fuels, the disposal of toxic wastes, the removal of rain forests and old-growth forests, increasing soil erosion and decreasing water quality due to intensive agriculture, the depletion of the earth's ozone layer, the reduction of biological diversity, and numerous issues dealing with population growth in our cities. Geographic information systems (GIS) have the potential to help us in our search for solutions to such problems.

A GIS can be defined as a system used for the (1) input, (2) storage, (3) manipulation, and (4) display of geographic data. The **data input subsystem** supports digitizing of analog map data or importing of digital spatial and image data. **Date storage and retrieval** functions provide for the storage and maintenance of both spatial and related nonspatial data files. **Data manipulation** programs perform analytical functions such as reclassification, overlay, and neighborhood analysis tasks. A **tabular and graphic output subsystem** is used for display of spatial and tabular data.

The software and hardware needed to run GIS with inexpensive personal computers is now widely available. Costs are continuing to decrease while computer capabilities are increasing. Consequently, the number of organizations using GIS is rapidly increasing in the United States and around the world.

Few textbooks on GIS are currently available. Moreover, the field of GIS is rapidly changing, with much of the material residing in the periodical literature. This volume contains an international collection of articles dealing with GIS concepts. The articles included were obtained primarily from refereed journals in cartography, GIS, and remote sensing. The compendium is designed to be useful in university GIS courses as a textbook supplement and as a reference for individuals interested in employing GIS technology. This book is the second GIS compendium published by the American Society for Photogrammetry and Remote Sensing (ASPRS) and the American Congress on Surveying and Mapping (ACSM). The first was published in 1987 and is entitled *Geographic Information Systems for Resource Management: A Compendium* (ISBN 0-937294-89-6). The main focus of the first volume was to primarily bring together articles on GIS applications, while the focus of this volume is on the fundamentals, principles, and issues in GIS. Particular attention has been given to such topical issues as entering digital data into a GIS, the assessment of errors in GIS, the development of improved spatial data structures to increase the efficiency of GIS, integration of remote sensing data, and the development of artificial intelligence techniques, and specifically expert systems, for GIS.

This book is divided into eight sections. Each section begins with a brief overview and a few suggestions for additional reading. The articles in the first section deal with the definition of GIS, the history of the technology, and requirements and principles for GIS implementation. The second section focuses on entry of spatial digital data into a GIS. Section three includes articles on the principles of logic as they apply to geographic data analysis, fundamental GIS operations, and digital terrain models. The fourth part of the book deals with GIS data quality issues and error assessment. Section five includes articles on the most recent developments and future trends in GIS including expert systems. Recent examples of GIS applications can be found in section six and information on how to select and evaluate a GIS is found in section seven. The final section provides a guide to information sources on GIS literature including books, journals, symposia proceedings, and an extensive selection of newsletters.

Acknowledgements are due to James W. Merchant for reviewing material and making suggestions on the design of this volume. Also I gratefully thank Donald F. Hemenway, Jr., ASPRS Director of Communications, for coordinating the technical aspects of publishing this volume. Finally, I thank the authors for contributing manuscripts and giving reprint permission to make this book possible.

William J. Ripple, Editor
Department of Forest Resources
Oregon State University
May, 1989

SECTION 1
Introduction

Overview

This introduction will provide the reader with background information on the definition, history, and general principles of geographic information systems (GIS) technology. Since there has been some disagreement regarding the definition of GIS, James R. Carter was invited to write the first paper on defining GIS. The second article by Parent and Church traces the history and evolution of GIS. Smith and others in the last article address some important issues related to the implementation of "large scale" GIS.

Suggested Additional Reading

Burrough, P.A. 1986. *Principles of Geographical Information Systems for Land Resource Assessment.* Oxford University Press, New York.

Dueker, Kenneth J. 1987. "Geographic Information Systems and Computer-Aided Mapping," *American Planning Association Journal.* 53(3):383–390.

ON DEFINING THE GEOGRAPHIC INFORMATION SYSTEM

Dr. James R. Carter
Associate Professor, Geography Department, and
Associate Director, Computing Center
University of Tennessee, Knoxville, TN 37996

The proliferation of computer systems with mapping capabilities poses a basis to question what it takes to make a geographic information system (GIS). There are installations and operations that everyone will agree are geographic information systems. There are other collections of hardware, software, data and personnel that may, or may not, be worthy of being designated geographic information systems, depending on one's concept of a GIS. While there is no authority to establish hard and fast definitions, it is incumbent upon each of us to develop our own concept and definition of a geographic information system. Most of us should have a flexible working definition of the geographic information system for use with such diverse groups as the general public, employers, managers, and professional colleagues.

There are many definitions of the geographic information system. Some definitions are implicit in the work of authors and presenters who tell what they have done, or can do, with their system which they refer to as a GIS. And, there are many explicit definitions of the geographic information system. Cowen, 1988, Goodchild, 1985, Marble, 1984, and Parker, 1987, in particular, have focused on definitions and capabilities of the geographic information system. Building on the discussion offered by Goodchild (1985), a rather universal and broad definition of the GIS might be 'an integrated system to capture, store, manage, analyze, and display information relative to concerns of a geographic nature.' With such a definition it can be argued that many organizations that systematically deal with maps and mapping can be said to have a GIS, whether the process involves the use of computers or not.

Of course, not everyone is content to accept such broad definitions without comment or qualification. I added the word 'integrated' to imply that there must be skilled and knowledgeable personnel and financial support in addition to the hardware, software and data. This point supports my contention (Carter, 1988) that the geographic information system is an institutional entity, reflecting an organizational structure that integrates technology with a database, expertise and continuing financial support over time. Thinking of geographic information systems in this way takes the GIS beyond being just a package of technology--hardware and software capable of performing GIS-types of functions. Accepting the GIS as an institutional entity, however, does not deny that the technology is an important part of any GIS as Parker (1987, 72) noted, "It is only when GIS is viewed as a technology, not simply a system, that its real significance becomes apparent."

Part of an Information System World

The use of the term 'Geographic Information System' in recent years is consistent with the convention of talking in terms of 'systems' and 'information systems' in many disciplines and areas of interest. It is common, for example, to talk about the welfare system, the food distribution system, and the education system. Trade journals and magazines use IS and MIS as part of a regular vocabulary, referring to information systems and management information systems. While much of this vocabulary is a matter of fashion, the need to better formalize procedures with the installation of complex computer systems has lead to a focus on what is information and how it can be managed in a systematic fashion.

One way of looking at the geographic information system is to see that it is only one of a variety of information systems. As such, the GIS community shares concerns with other information system communities relative to 1) the organization of the information in such a fashion that it will have utility when retrieved, 2) the management of the system so that appropriate information is available to those who need it and appropriate access is limited to those who should not be able to see or to modify data, 3) the continued support and maintenance of the information system over time so that data are current and users have confidence in the data, 4) the continuing need to upgrade the system to keep it at current levels of technology, 5) the provision of backups and redundant systems, 6) the need to build and retain a staff, 7) the need to educate users and management as to the realistic expectations of the system, and 8) the tendency to oversell the system at the beginning and to create expectations that are difficult or impossible to meet.

Subdividing the Generic Field

Relative to the broad range of information systems, the geographic information system seems to be a very specialized and focused area of concern. However, those of us within the GIS community know there is great diversity within our midst. This suggests that there is a need for at least two types of definitions: one for the world at large who know little or nothing about GIS and another for those within the GIS communities who are seeking to make some sense out of all that is going on in the world of GIS (Jeyanandan, 1988, 169-70). Most basic definitions, such as that given at the beginning of this paper, should suffice for the public at large. However, no single statement will probably ever suffice for internal consumption, for within the generic area of GIS there are natural divisions emerging to the point that we see the use of other terms to define specific types of geographic information systems.

In some cases users substitute another word or expression for 'geographic' to give such things as the Urban Information System or the Natural Resource Information System. On the other hand, there are other types of geographic information systems that are so distinct from the generic GIS that the personnel working in these systems tend to operate in their own communities of interest. The AM/FM or Automated Mapping/Facilities Management community is primarily concerned with utilities and the built environment. This community now has its own organizations and literature. Another community that might be thought of as standing apart from the generic GIS is that community focused on the land

4

parcel, or as it is more commonly referred to in the international sphere, the cadastre (Kjerne and Dueker, 1986, 142-43). Many persons now refer to such a system as a Land Information System (LIS), or in some cases a Multipurpose Cadastre (MPC). For some professionals the scale of investigation and the specific focus of the LIS community are distinct enough to stand apart from the more generic GIS, as Kjerne and Dueker (1986, 143) note ". . . a land information system is the application of geographic information systems technology to land records data."

Further, I contend that there are also many systems that are functionally geographic information systems but that are not thought of as geographic information systems (Carter, 1988). Among these systems are weather forecasting systems, marine exploration and navigation systems, petroleum exploration systems, and transportation routing and modelling systems. There is no universal geographic information system, nor should we ever expect to see one, for there is little commonality between the variety of tasks that such systems are designed to address. Instead, we see a range of geographic information systems organized around functional tasks.

Cores and the Periphery

At the heart of the GIS community this author observes two cores of emphasis. One core consists of those persons and organizations charged with building and maintaining a geographic database relative to day-to-day operations. The prime focus of such persons must be on the integrity of the database. Frequently in such organizations there is little opportunity or cause to push the frontiers with expanded functionality. AM/FM systems in some urban areas and large systems employed by firms that manage extensive tracts of resources are probably the best examples of this type of GIS community. The other core focuses on capability and expertise. The persons in this group are seldom concerned with the continuing process of updating a database and maintaining its integrity. Rather, this group is primarily concerned with techniques and procedures of digitizing, data structures, analysis, image processing, modelling, formats for display, and user interfaces. Within this core of emphasis, databases are project specific. University laboratories and some service bureaus are probably the best examples of this community. In terms of geographical data handling needs, Tomlinson and Boyle (1981, 65) observed that users could be thought of as either 'project related' or 'inventory related;' a distinction not dissimilar to the two cores identified here.

Clustered around these two cores are communities of support. Hardware and software vendors provide basic tools and often are the major conduits for the interchange of ideas between the two cores. Trainers, consultants, and the professional associations add another dimension as they facilitate the interchange of knowledge of such systems. Land surveyors, census officials, foresters, soil scientists, cartographers, photogrammetrists, remote sensing specialists, et.al. provide and use the databases that form integral parts of specific GIS's. These experts also bring knowledge about the limitations in their databases and for this reason these persons may be considered a part of the GIS community. Academics are charged to produce a cadre of persons with a foundation of knowledge about the GIS and/or allied disciplines and are often in a position to monitor objectively the developments in the disciplines and to expand and refine the possibilities of geographic information systems.

Scales of Focus

Scales of investigation and sources of data are still another basis for differentiation within the full realm of GIS. Some persons define geographic information systems as those information systems concerned with spatial topics--spatial information systems (Cowen, 1988, 1554, and Parker, 1987, 73). While the GIS has to have a spatial dimension, there are many spatial concerns that are not geographic. Astronomers chart the heavens while the medical community images the body and molecular biologists focus on DNA structure. To say the GIS is a spatial information system is only partly correct. To say a spatial information system is a GIS is too cavalier. The term 'geographic' defines the range of scales of spatial investigation from the globe at one extreme to the land parcel at the other. Thus, the GIS may be world-wide in scope, specific to a community or at any range of scales in between. Frank (1988, 1557-58) delimits this range of scale as
". . . data related to location in real world space."

Data appropriate to one scale of investigation seldom have utility for other scales of investigation. Further, the sources of data vary as a function of scale. Detailed data for local investigations generally have to be gathered locally or contracted for by the organization building the GIS. By contrast, regional and global data normally are collected by agencies and organizations independent of those building the GIS. Therefore, the GIS communities with the local, large-scale focus employ different standards and techniques to collect and evaluate their databases than do the GIS communities focused on regional and global scale questions.

Conclusions

The diversity of the GIS community is as great as the diversity of communities built around map makers and map users. There is information in a world map of soil types and information in a soils map of a farm. Both maps deal with soils. It takes some knowledge of soils to appreciate the information contained in each map. While these maps have a common link in that they give a spatial representation of some aspect of the soil, the maps have little in common in terms of the ways the data were collected, the construction of the maps, and the audience who can use the maps. Comparable communities exist in the GIS realm.

Building on consideration of the many dimensions noted above leads me to a working definition of the geographic information system as 'any of those information systems that focus on spatial concerns and phenomena in the range of scales from the whole earth to the land parcel. Geographic information systems share the concerns of all information systems, with the added need to accommodate the spatial/temporal dimensions. There are a variety of geographic information systems, many of which are known by other names.'

References

Carter, James R., 1988, A Typology of Geographic Information Systems, *Technical Papers, 1988 ACSM-ASPRS Annual Convention*, Vol. 5, 207-215.

Cowen, David J., 1988, GIS versus CAD versus DBMS: What Are the Differences?, *Photogrammetric Engineering and Remote Sensing*, Vol. LIV, No. 11, Nov. 1988, 1551-55.

Frank, Andrew U., 1988, Requirements for a Database Management System for a GIS, *Photogrammetric Engineering and Remote Sensing*, Vol. LIV, No. 11, Nov. 1988, 1557-64.

Goodchild, Michael F., 1985, Geographic Information Systems in Undergraduate Geography: A Contemporary Dilemma, *The Operational Geographer*, No. 8, 34-38.

Jeyanandan, D., 1988, A Systems Approach to Land Information Systems, *Surveying and Mapping*, Vol. 48, No. 3, Sept. 1988, 161-71.

Kjerne, Daniel and Kenneth J. Dueker, 1986, Modeling Cadastral Spatial Relationships Using an Object-Oriented Language, *Proceedings, Second International Symposium on Spatial Data Handling*. Williamsville, NY: International Geographical Union, 142-57.

Marble, Duane F., 1984, Geographic Information Systems: An Overview, *Pecora 9 Proceedings: Spatial Information Technologies for Remote Sensing Today and Tomorrow*, Sioux Falls, SD, 18-24, reprinted in William J. Ripple (ed.), *Geographic Information Systems for Resource Management: A Compendium*. Falls Church, VA: American Society for Photogrammetry and Remote Sensing and American Congress on Surveying and Mapping, 2-8.

Parker, H. Dennison, 1987, What is a Geographic Information System? *GIS '87 - San Francisco*. Falls Church, VA: Amercian Society of Photgrammetry and Remote Sensing and American Congress on Surveying and Mapping, 72-80.

Tomlinson, R. F. and A. R. Boyle, 1981, The State of Development of Systems for Handling Natural Resources Inventory Data, *Cartographica*, 18(4): 65-95, reprinted in William J. Ripple (ed.), *Geographic Information Systems for Resource Management: A Compendium*. Falls Church, VA: American Society for Photogrammetry and Remote Sensing and American Congress on Surveying and Mapping, 34-63.

Paper Presented at the GIS '87 Conference, San Francisco, October, 1987

EVOLUTION OF GEOGRAPHIC INFORMATION SYSTEMS AS DECISION MAKING TOOLS

Phillip Parent & Richard Church
Department of Geography
University of California
Santa Barbara, CA 93106

ABSTRACT

Throughout the history of the development of geographic information systems (GIS) the primary goal has been to take raw data and transform it, via overlays and other analytical operations, into new information which can support the decision making process. Geographic information systems have their roots in the history of thematic cartography. Modern GIS evolved from the combination of increased computational capabilities (the computer revolution), refined analytical techniques (the quantitative revolution), and a renewed interest in environmental/social responsibility. The earliest systems, such as the Canadian Geographic Information System (CGIS), Storage and Retrieval of Water Quality Control (STORET), and Map Information Assembly and Display System (MIADS), were all applications oriented. They were developed to facilitate policy decisions by public agencies with specific mandates. We examine the roots and origins of GIS, trace it's development as a decision making tool, and examine the features that differentiate GIS from an automated inventory device.

INTRODUCTION

The geographic information system community is a diverse crowd. Foresters, geographers, municipal governments, utilities, and academic researchers all are exploring the same territory, but from different directions. The binding tie is that they all deal with spatially referenced data, that is attributes of a specific area on the Earth's surface. Each application requires the reduction of many different 'layers' of data into a new layer of information, usually in cartographic format, that can support the decision making process. The key feature of a GIS is the analysis of data to produce new information. This information can consist of simple overlays to complex simulations.

This broad view of GIS encompasses many disciplines and applications. A GIS is not by definition a computer-based system, although it is generally considered to be such. GIS is not a recent invention. By looking at the historical roots of GIS, we can trace certain themes throughout the last 200 years that have led us to the present technology. GIS has the potential to change the way we look at the world. A knowledge of the background and history of GIS is essential in order to help guide it's development.

ROOTS OF GIS

The roots of geographic information systems date back to the mid-eighteenth century with the development of cartography and the production of the first accurate base maps. Until that point, the graphic depiction of spatial attributes could not be accurately portrayed. Thematic mapping, that is the definition of attributes to a given geographic area, came about soon after, with maps portraying magnetic variation with isolines and wind direction by means of arrows (Robinson, 1982). The idea of recording different layers of data on a series of similar base maps was an established cartographic convention by the time of the American Revolutionary War (Harley et al, 1974). In fact, a map of the siege of Yorktown done by the French cartographer Louis-Alexandre Berthier contained hinged overlays to show the troop movements in the final battle (Rice & Brown, 1974).

The refinement of lithographic techniques enabled the thematic cartographer to become more precise in data representation. More accurate information could be shown and clearly reproduced. These technical advancements gave scientists the tools with which they could carry on their research.

Advances in the physical and social sciences provided geographers with the intellectual tools for spatial analysis. Statistical techniques, number theory, and advanced mathematics were all beginning to be developed. The first geologic map of Paris appeared in 1811 followed by one of London in 1815. The work of Alexander Von Humboldt was influential in both areas. The British Census of 1825 produced a tremendous amount of data that had to be analyzed. The science of Demography soon evolved.

The late 18th century also saw an increase in social awareness. Publication of *The Wealth of Nations* by Adam Smith in 1776 and *An Essay on the Principle of Population* by Malthus in 1798 had a profound effect on social thought. Resources were beginning to be recognized as a finite entity. The age of social responsibility was dawning.

By 1835 technology (advanced cartographic techniques), science (social science theory), and social thought (environmental responsibility) had advanced to the point that the combination of these three factors could support more comprehensive thematic mapping projects. But it was economics, specifically the industrial revolution, that was the main catalyst in the evolution of GIS. The explosion in manufacturing increased the need for raw resources, drew people into crowded urban environments, and created the need for an extensive infrastructure, both social and industrial, especially in the transportation field. In fact it was a transportation study completed in 1837 that first brought together the technical, social, and scientific advancements of the previous decades into an integrated whole.

The Atlas to Accompany the Second Report of the Irish Railway Commissioners which appeared in 1838 was perhaps the first Geographic Information System per se. The atlas consisted of a series of maps that depicted population, traffic flows, geology, and topography. For each sheet the base map was uniform in regard to scale and county boundaries (Robinson, 1955). By mentally 'overlaying' the different attributes at a given spatial location, the commissioners could make their recommendations as to where the best transportation routes could be sited. In effect, this series of maps was a decision-making tool.

Thematic mapping continued to improve with the London publication in 1848 of the first truly uniform world atlas in respect to scale, symbolization, and cohesiveness. Such conventions as isobars, graduated circles, and volume lines were used extensively. However, this atlas was for general use rather than a specific application like the Railway Atlas. Thematic maps continued to develop in scope and purpose. The famous Snow map of 1854 effectively solved a public health crisis by pinpointing a contaminated water-well as a source of cholera during a London epidemic. By representing attributes (in this case sickness) in a geographical manner, decision-making was improved.

The years of 1835 - 1855, the "Golden Age" of cartography, saw many innovative techniques. Although there were no series of breakthroughs in the following 100 years, techniques improved steadily. The introduction of Hollerith punch cards and machine processing in the 1890's was a harbinger of things to come but had little impact on cartography. This situation remained relatively stable until well into the 20th century.

EARLY MODERN SYSTEMS

The three factors which acted as catalysts in the breakthroughs of the 1830's and 1840's, technology, science, and social awareness, re-exerted themselves in the mid-20th century. The first, technology, manifested itself in the late 1940's with the development of the first electronic computers. By 1952 ENIAC was processing the 1950 Census data. The process of electronic calculation opened new possibilities of research based on the massive manipulation of huge data files. Intricate models and alternatives could be generated to simulate future events.

At first, output was limited to alpha-numeric characters in a non-spatial representation. But by the late 50's meteorologists, geophysicists, and geologists had incorporated computer-generated maps into their work. Graphic representation was very crude in the early instances. Output was derived from different combinations of alpha-numeric characters which showed different shades. However crude, at least the capability to print out geographic data in a cartographic format had been established.

The science of spatial analysis was also rapidly advancing at the same time. In Sweden, Dr. Torsten Hagerstrand was developing his theory of spatial diffusion. By the early 1960's, modelling was being applied to census data by both academics and government in the United Kingdom (Rhind, 1981), but it was not widespread. In the United States at the same time urban data banks were being established to help implement data processing for state and urban governments (Kraemer, 1969). The tools for innovative geographic processing, both technical and theoretic, had advanced greatly in the period from 1945 - 1965.

The final factor, social awareness, also underwent a metamorphosis in the same time-frame. A renewed interest in the environment, fueled in part by the 1962 publication of Rachel Carson's book *Silent Spring*, combined with the realization of the need for improved urban planning, gave impetus to social scientists to refine their geographic techniques (Estes, 1986). Resource inventory and the modelling of physical processes were becoming more and more common. By 1962 local and state governments had started showing great interest in geographic information processing as a decision-making tool (Kraemer, 1969).

By the mid-twentieth century, the three major factors for cartographic change, technology, theory, and social awareness, had each evolved to an advanced state. The critical mix of digital computers, improved analytical techniques, and an increased awareness of the finite nature of the Earth's resources had set the stage for the development of the first modern Geographic Information System.

As in Great Britain 80 years earlier, it was a socio-economic need, transportation studies, that really spawned the first electronic geographic information system. In 1955 the Detroit Metropolitan Area commissioned a traffic study to plan for future needs. The study group divided Detroit into an grid of one-quarter square-mile cells and then inventoried the traffic flow through each cell. Then, by using statistical analysis, they predicted future traffic volumes. This future model of traffic was then used to site and prioritize future highway development (Detroit Metropolitan Area, 1955). Although this study used various innovative techniques, no computer output was in a geographically ordered format. Four years later, however, the Chicago Area commissioned a similar study, only this time using computers to a much greater extent. In fact, an actual machine, the Cartographatron, was developed by the Armour Research Foundation to graphically portray the volume of traffic over certain routes. This output was then incorporated into the final report for transport analysis (Chicago Area Study, 1959).

This transportation modelling study might well have been the first actual GIS application. It was a decision making tool in the sense that the system created new information by the statistical analysis of different data layers and displayed the output on a computer graphics device in a format that fostered effective decision making.

By the early 1960's, academicians were beginning to undertake research along these same lines. An early center of such computer-assisted quantitative analysis was the University of Washington at Seattle. Under the tutelage of Dr. William Garrison in the Department of Geography, graduate students studied advanced statistical methods, rudimentary computer programming, and applications in transport modelling, among other things (Morrill, 1983). The 1959 report *Studies of Highway Development and Geographic Change*, a seminal work on transport analysis, listed as it's authors Garrison, Brian Berry, Duane Marble, Richard Morrill, and John Nystuen. Other students of that era included Michael Dacey, Art Getis, and Waldo Tobler. Later, in 1962, a group working under Dr. Edgar Horwood of the Department of Civil Engineering developed the ROMTRAM system, an early GIS that incorporated a computer-generated statistical mapping technique for the Department of Housing and Urban Development which would identify targets for urban renewal (Gaits, 1969). The University of Washington was perhaps the first academic locus for the study of geographic data processing. The happy combination of theory, technique, and vision that existed then has left it's mark on the GIS community to this day.

If transportation studies were the first automated geographical analysis systems with statistical capabilities, resource inventories gave widespread applications to the emerging field. Governments in Canada and the United States began to see the opportunities that could be realized by the electronic storage and retrieval of land based data. In 1962 Roger Tomlinson of the Canada Land Inventory developed the Canadian Geographic Information System (CGIS), the first system to be called a geographic information system although it wasn't named as such officially until 1965 (Tomlinson, 1976). Unlike the systems mentioned above, CGIS was designed for more than just one specific application. It's major application was to store digitized map data and land-based attributes in an easily accessible format for all of Canada. A polygon-based system, it had significant shortcomings in interactive capabilities to the point that real time graphic editing was severely limited. CGIS had storage and retrieval capabilities plus it could re-classify attributes, change scales, merge and create new polygons, and create lists and reports. The system is still in operation, actively digitizing data and generating output for a multitude of applications (Tomlinson, 1984). This system, then, according to the definition, is a true GIS, a system that can create new information layers and generate graphic output to support the decision-making process.

In the United States, the Division of Water Supply and Pollution Control of the Public Health Service developed the Storage and Retrieval of Data for Water Quality Control System, or STORET. It was designed to standardize the data generated by different governmental and private agencies in regards to hydrological characteristics, ie. water quality, flows, treatment processes, and location. Data, in the form of intricate codes, was entered manually on punch cards and fed into the computer. The data-base could then be queried for specific information. Output was generated in tabular form. There was no provision at all for graphic representation (Green, 1964). Supplementary statistical packages could be utilized for data analysis, but that component was not an integral part of the system.

At the same time the Forest Service was developing it's own land management assembly and display system, MIADS, at Berkeley. This was a much more advanced system, one that could not only store and retrieve attributes of a given grid-cell, but perform simple overlay functions and mathematical calculations and prepare simulations over time. Output was generated on a line printer, but was arranged topologically to produce a cartographic image. In addition, updating data was a fairly straightforward task (Amidan, 1964).

MIADS, then, might be considered the first U.S. 'full-service' GIS in the natural resource environment. The system was designed to store data, manipulate and analyze the data to create new information layers, and then to generate output in a graphic format. Very soon after other governmental agencies, federal, state, & local, began to develop their own systems that ranged from simple automatic inventories to interactive systems with advanced analytical capabilities. The key attribute of these systems is analytical processes. Let's take a closer look at modelling and it's development in the GIS context.

GIS AND ENVIRONMENTAL INVENTORY AND MANAGEMENT

In 1925 Streeter and Phelps published a report on the effects of pollution and natural purification on the Ohio River (Streeter & Phelps, 1925). They developed and presented a first order function that related water quality (i.e. dissolved oxygen) as a function of the amount of sewage discharge, location of discharges, and the condition of the stream (e.g. flow rates, depth, reaeration rate, etc.). Although this equation has been expanded in different ways, their basic process model is still used today to model water quality along a stream. With the quantitative understanding of the impacts on water quality based on a geographical distribution of pollutant sources, it became obvious that management strategies needed to be expanded in scope and size to encompass entire watersheds. This basic understanding along with the concern for quantity and use issues led to the development of a number of river basin management authorities such as the Ohio River Basin Sanitation Commission. With the availability of the computer and operations research tools in the 1960's, researchers began to propose water quality management strategies for entire river basins based on linear and dynamic programming and process models like Streeter-Phelps (see Liebman (1965) and Revelle, Loucks, and Lynn (1967)). Studies included major river systems such as the Potomac, Delaware, and Willamette Rivers. Such models required large scale geographic data, spatial relationships, and quantitative models of environmental processes and alogorithms. Clearly, such models are specialized types of GIS where data information and retrieval were less important than areal management strategy identification and impacts analysis. Today it is common for many regions or metropolitan areas to have complex water resource system models that are operated with large geographic data bases. Much of this development was funded by the U.S. EPA under the 208 basin-wide planning program established under public law 92-500 in 1970. An excellent example is the HEC-SAM model that was developed by the U.S. Army Corps of Engineers at the Hydrological Engineering Center in Davis, California. The HEC-SAM model utilizes both grid and polygon based data and was designed to provide the capability to assess hydrologic, flood damage, and environmental consequences of development reflected by alternative land use patterns and management policies. In addition to the early STORET system of EPA, the USGS has operated two large and specialized water resource GIS systems called REACH and WATSTORE. Such systems will continue to be enhanced and perhaps integrated with other large systems so that more users can be accommodated and are able to easily access such data.

With the development of good atmospheric dispersion models in the early 1960's for small airsheds (see Turner (1964)), it became possible to integrate areal emissions data and either predict air quality in the air shed or determine optimal policies for meeting air quality standards. For example, Teller (1968) developed and applied a model for St. Louis that found the optimal minimum cost policy of fuel purchases for industrial boilers so that sulphur oxide air standards could be met. This model required significant spatial data. Unfortunately, the use of many of the early decision assisting models relied on ad hoc developed data structures, were developed for a single purpose, and lacked program support for continued development and data base management. However, what they did demonstrate was that large scale system models were capable of integrating a large amount of data, solving complex problems, and produce results within reasonable amounts of effort. The benefits of these models were hard to quantify and could not justify (each on an independent basis) their own data system with the costs of continued data collection and maintenance. Further, hardware costs at that time kept storage and computation costs relatively high. It is clear that this early evolution flourished with possibilities but faded by the pragmatics of costs, hardware, lack of common data bases, and good algorithms. Now, the outlook has changed with less expensive equipment, more efficient data structures and algorithms, automatic monitoring, and digital conversion of environmental data.

The development of GIS has not been one without large failures. In beginning the development of any large scale data base, the uses, users, and benefits should be identified. Further alternatives should be explored and evaluated. A decision to develop must be based on realistic expectations and costs. Those systems that were not developed with a prepared plan had a good chance of failing. In 1966 Governor Nelson Rockefeller of New York pledged that there would be a natural resource inventory of the state. This pledge was made without a clear plan other than the idea that such an inventory would be important and valuable. Based on this pledge, The Land Use and Natural Resource (LUNR) inventory system was developed. LUNR, produced by researchers at Cornell University, was based on a cell size of 1 square kilometer and covered the entire state. Begun in 1967 and completed in 1970, the data base contained information on 130 categories of land use. The system was capable of producing inventory maps and overlays. Although the LUNR system cost more than $750,000 to capture data and develop, the state allocated only $60,000 per year for improvements and maintenance. Early user's greatest complaint was that the data was out of date (Power, 1975). Obviously the fate of the system was determined before it was even tested. The uses and users were not well identified and as such the system did not satisfy any specific need. The history of LUNR probably made a number of state agencies in New York keep GIS at arms length until benefits, costs, and needs could be better defined. If the development of a system is not based on a realistic plan with realistic goals, then there is a good chance of failure.

Much of the early development of GIS systems was dictated by a specific need rather than by a set of different types of needs. Data base collection and development was typically chosen in order to match the specific need as much as possible, keep costs of collection low, and fit within the constraints of available computer equipment and hardware. As an example, grid based systems became simple but popular approaches to inventory large scale resource data. For example, MIADS, GRIDS, MAGI and early versions of ORRMIS (to name a few) all used cell based structures. Basic functions varied from simple inventory (eg. GRIDS) to sophisticated regional modelling as in ORRMIS. Although most of the ORRMIS models were defined for grid data, they could handle and store polygon data. The system transformed polygon information into cell information whenever any of the models were applied (Durfee, 1974). Further, a company that wanted to enter the field with a model usually found it more expedient to develop a cell based system. An example of this was the GIMS system of Dames and Moore (1975) that was used to do overlay analysis and applied to various siting problems. In the mid 1970's a corporate tool like a simple cell based overlay program was an important asset when it came to selling consulting services.

Of course the use of polygon and other data structures followed primarily because of the inability to handle boundaries and points in a cell structure. The type of equipment that was needed was substantially more sophisticated and the actual costs of operation increased as well, whenever actual maps were produced or digitized. Today it is fortunate that the costs of peripherals has decreased to the point that this technology can be easily used and implemented at modest cost.

GIS AND URBAN MODELLING

Along with the development of environmentally based GIS models in the 1960's there was a similar development of large-scale geobased urban models. The first generation of urban simulation models held 'great promise' but in fact did not progress to the point of being useful or accurate. Lee (1973) states that the characteristics of these models were: 1) they were large in the sense that the only practical way to operate them was by utilizing the computer; 2) commonly they were spatially disaggregated and allocated activities to geographic zones; 3) they pertain to a single specific metropolitan area. By and large one can classify these models as GIS based systems, where again the main feature was to take spatial data and simulate a process over time in order to assist in decision making. Unfortunately (or maybe fortunately) many of these first generation models were discarded due to a number of problems including cost, misspecification of human behavior, error and complexity issues, and the lack of a good evaluation procedure. Lee characterized this first generation as dinosaurs that collapsed rather that evolved.

By the early 1970's a second generation of urban models was under development. However, the goals were far less robust and the models were usually simpler. To this day, skepticism in large scale simulation exists based on the failure of the early urban GIS simulation systems that were designed to 'model' rather than keep track of inventory. If it had not been for the consistent growing long term need to keep a basic inventory of urban land, utilities, and property records, GIS development applied to urban areas would have substantially cooled. With the consistent improvement of new GIS systems with increasing capabilities, many of the long term inventory and mapping needs have been met. With this success and gained credibility, researchers can now return to this basic modelling area with the chance to finally approach the goals of those that were identified over twenty years ago.

One of the second generation models that was developed in Australia in 1970 was TOPAZ (Technique for the Optimum Placement of Activities in Zones). TOPAZ is a general planning technique that has been applied at the regional, urban, and facility planning levels. The TOPAZ model has met with continued success as it has evolved over the last fifteen years. In one documented study in 1980, TOPAZ had been applied to 40 different areas in the USA and Australia (Brotchie, 1980). The model has been used in conjunction with urban geobased data to analyze sewer and water expansion and costs, trip generation and mode choice, water catchment planning, investigating issues originating with corridor developments and new communities, and planning of future urban services. Because the system can allocate activities to spatial zones, it has also been used to lay out buildings and complexes as a micro-scale GIS. Models like TOPAZ can be classified as decision assisting techniques and are substantially different than special real time models like that used in computer aided dispatching for emergencies.

One of the early systems that has had a great impact in applications is the Dual Independent Map Encoded (DIME) file system of the Census Bureau. The DIME file represented a dual encoded map of streets and intersec- tions along with geographic codes and addresses. Each street segment is coded twice, once as an edge to one block (right) and once as an edge to the other block (left). Geographic coordinates are specified for each intersection. With each node and block (a unique area defined by a boundary of street segments) uniquely numbered, the computer can then construct two independent networks and match them to insure that all areas are accounted for and that the network is complete (Totschek, 1969). Although technically the DIME file is only a data file, it has probably been used in more applications than any other system.

With the new TIGER system (Bureau of Census) and special commercial data files like that of ETAK (currently being coded for most metropolitan areas of the US), sophisticated applications in navigation, vehicle routing, scheduling, dispatching, and automatic vehicle location (AVL) will become commonplace. Most cities above 100,000 population now have computer-aided dispatching systems that are keyed to a geo-base of street records. With an address from a call or the ATT 911 emergency system, an incident's exact location can be identified within a second of wall clock time. Some of the geo-base files are sophisticated structures that allow an efficient address search and verification in a large metropolitan areas such as Los Angeles. Within the next five years, many city emergency services will be capable of displaying on a graphics terminal (in real time) a map with exact positions of all types of equipment including fire, police, and ambulance automatically. The use of GIS in cities will expand beyond property records and planning functions to many real time applications as well. As an example, the City of San Diego has plans to integrate over 27 major city operations on one system. These include such functions as school districting and busing, water and sewer operations, and planning (F-M Automation Newsletter, 1987).

REFERENCES

Amidon, Elliot L., 1964, *A Computer-Oriented System for Assembling and Displaying Land Management Information*, U.S. Forest Service Research Paper PSW-17, Berkeley, CA., Pacific SW Forest & Range Experiment Station.

Brotchie, John F., Dickey, John W., and Sharpe, Ron, 1980, *TOPAZ*, Lecture Notes in Economics and Mathematical Systems, Vol. 180, Berlin, Springer-Verlag.

Chicago Area Transportation Study, 1959, *Final Report in Three Volumes*, Volume 1, Survey Findings, Chicago, IL

Dames & Moore, 1975, *Data Management for Power Plant Siting: Delmarva Interface Study*, Cranford, NJ, Dames & Moore.

Detroit Metropolitan Area, 1955, *Traffic Study*, Detroit, MI

Durfee, R.C., 1974, *ORRMIS - Oak Ridge Regional Modeling Information System*, Oak Ridge, TN, Oak Ridge National Laboratory.

Estes, John E., 1986, "A Perspective on the Use of Geographic Information Systems for Environmental Protection", in *Conference on GIS for Environmental Protection*, Las Vegas, NV.

F-M Automation Newsletter, July, 1987, "City & County of San Diego: Going With IBM?", Vol. 11, No. 7, pp. 1 - 7.

Gaits, G.M., 1969, "Thematic Mapping by Computer", *Cartographic Journal*, Vol. 6, No. 1, pp. 50 - 68.

Green, Richard, 1964, *The Storage and Retrieval of Data for Water Quality Control*, Public Health Service Publication No. 1263, Washington, DC, U.S. Department of Health, Education, & Welfare, Public Health Service.

Harley, John B., Petchenik, Barbara B., and Towner, Lawrence W., 1978, *Mapping the American Revolutionary War*, Chicago, IL, University of Chicago Press.

Kraemer, Kenneth L., 1969, "The Evolution of Information Systems for Urban Administration", *Public Administration Review*, July/August, pp. 389 - 402.

Lee, Douglass B., 1973, "Requiem for Large-Scale Models", *Journal of the American Institute of Planners*, Vol. 39, No. 3, pp. 163 - 178.

Liebman, J.C., 1965, *The Optimal Allocation of Stream Dissolved Oxygen Resources*, Doctoral Thesis, Ithaca, NY, Cornell University.

Morrill, Richard, 1983, "Recollections of the 'Quantitative Revolution's' Early Years: The University of Washington, 1955-65", in *Recollections of a Revolution: Geography as a Spatial Science*, Billinge, Gregory, and Martin, eds., New York, St. Martin's Press, pp. 57 - 72.

Power, Margaret 1975, *Computerized Geographic Information Systems: An Assessment of Important Factors In Their Design, Operation, and Success*, Master's Thesis, Saint Louis, MO, Washington University, Sever Institute of

Technology.

Revelle, C., Loucks, D., and Lynn, W.R., 1967, "A Management Model for Water Quality Control", *Journal of Water Pollution Control Federation*, Vol. 1, No. 4, pp. 477.

Rhind, David, 1977, "Computer-Aided Cartography", *Transactions of the Institute of British Geographers*, Vol. 2 (new series), No. 1, pp. 71 - 97.

Rhind, David, 1981, "Geographical Information Systems in Britain", in *Quantitative Geography: A British View*, Wrigley and Bennet, eds., London, Routledge and Kegan Paul, pp. 17 -35.

Rice, Howard C. Jr. and Brown, Anne S.K. (eds. and translators), 1972, *The American Campaigns of Rochambeau's Army*, Princeton, NJ, Princeton University Press.

Robinson, Arthur H., 1955, "The 1837 Maps of Henry Drury Harness", *Geographical Journal*, Vol. 121, pp. 440 - 450.

Robinson, Arthur H., 1982, *Early Thematic Mapping in the History of Cartography*, Chicago, IL, University of Chicago Press.

Streeter, H.W. and Phelps, Earle B., 1925, *A Study of the Pollution and Natural Purification of the Ohio River*, Public Health Bulletin No. 146, Washington, DC, Public Health Service (reprinted 1958).

Teller, A., 1968, "The Use of Linear Programming to Estimate the Cost of Some Alternative Air Pollution Abatement Policies", in *Proceedings of the IBM Scientific Computing Symposium on Air and Water Resource Management, White Plains, NY, IBM, pp. 345 - 353*.

Tomlinson, Roger, 1976, *Computer Handling of Geographical Data*, Paris, The UNESCO Press

Tomlinson, Roger, 1984, "Geographic Information Systems - A New Frontier", in *International Symposium on Spatial Data Handling*, Keynote Address, Zurich, Switzerland.

Totschek, Bob, Almendinger, Vlad, and Needham, Ken, 1969, *A Geographic Base File for Urban Data Systems*, Santa Monica, CA, System Development Corp.

Turner, D.B., 1964, "A Diffusion Model for an Urban Area", *Journal of Applied Meteorology*, Vol. 3, No. 1, pp. 85 - 91.

Requirements and principles for the implementation and construction of large-scale geographic information systems

TERENCE R. SMITH

Department of Computer Science, University of California at Santa Barbara, California, U.S.A.

and SUDHAKAR MENON, JEFFREY L. STAR and JOHN E. ESTES

Department of Geography, University of California at Santa Barbara, California, U.S.A.

Abstract. This paper provides a brief survey of the history, structure and functions of 'traditional' geographic information systems (GIS), and then suggests a set of requirements that large-scale GIS should satisfy, together with a set of principles for their satisfaction. These principles, which include the systematic application of techniques from several sub-fields of computer science to the design and implementation of GIS and the integration of techniques from computer vision and image processing into standard GIS technology, are discussed in some detail. In particular, the paper provides a detailed discussion of questions relating to appropriate data models, data structures and computational procedures for the efficient storage, retrieval and analysis of spatially-indexed data.

1. Introduction

This paper presents an overview of the structure of 'traditional' geographic information systems (GIS) and a discussion of several important issues concerning the design, implementation and application of large-scale GIS in relation to current computer technology. In its simplest form, a GIS may be viewed as a data base system in which most of the data are spatially indexed, and upon which a set of procedures operates in order to answer queries about spatial entities in the data base. This simple form, however, belies many of the extremely difficult and complex problems that confront researchers and practitioners who wish to construct large-scale GIS.

Many of these difficult problems arise because environmental scientists, resource managers and other users of spatial data are acquiring access to very large and rapidly increasing volumes of spatially-indexed data. The key to the efficient use of such data sets is the existence of powerful systems capable of acquiring data from a variety of sources; changing the data into a variety of useful formats; storing the data; retrieving and manipulating the data for analysis; and then generating the outputs required by a given user. The design and implementation of GIS that are capable of efficiently performing these basic functions on large volumes of spatially addressed data are therefore of critical importance.

The paper is structured as follows. In the remainder of this section, we present a brief history of GIS development and list the five major components of a GIS. We then discuss the requirements that a large-scale GIS should satisfy, as well as listing a set of basic principles that may be used in designing and implementing GIS that satisfy such requirements. In the next major section, we discuss 'traditional' GIS, in terms of both the five major components and their applications. Then follows an examination of a number of key issues that are important in the design and implementation of large-scale

GIS, including the application of systematic knowledge from several sub-fields of computer science, as well as discussing GIS that integrate image-processing capabilities and the effects of hardware developments on GIS design and use.

In particular, we focus most of our attention on questions relating to the choice of spatial data models and data structures. The reason for this concentration is that by far the greatest difficulties of designing, implementing and using GIS arise from the fact that their essential characteristic involves the processing of large volumes of spatially-addressed data. It is therefore essential that great attention be focused on both the data structures that are used to store geographic information and the efficiency of the procedures that are used in relation to these data structures to store, analyse and retrieve data.

1.1. *A brief history of GIS*

Technological advances in computation, cartography and photogrammetry in the 1940s and 1950s laid the foundations for the development of GIS, which began in the 1960s. The conceptual framework within which early GIS were implemented involved individuals from many disciplines, who realized the need for computers in integrating data from a variety of sources, in manipulating and analysing the data and in providing output which could be used as part of a decision-making process.

The Canadian Geographic Information System was implemented in 1964, one year after the first conference on Urban Planning Information Systems and Programs. This conference led to the establishment of the Urban and Regional Information Systems Association. The New York Landuse and Natural Resources Information System was implemented in 1967, and the Minnesota Land Management Information System in 1969. In these early years, the costs and technical difficulties of implementing full-scale GIS systems were such that only large users of geographic information (such as federal and state agencies) could afford their development. McHarg's influential book (1969) formalized the concept of land suitability/capability analysis, and was important in influencing further work in the use of overlays of spatially-indexed data in resource planning and management decision making.

There has since been a rapid increase in the number of GIS, as a result of both advances in computer technology and increases in the availability of spatially-referenced data in digital form. A United States Department of the Interior, Fish and Wildlife Service report published in 1977 compared the selected operational capabilities of 54 different GIS in the United States (USFWS 1977). This survey, which is representative of several others conducted in the late 1970s, provided information on hardware environments, operating environments, programming languages, data acquisition, data types, data bases and documentation.

To date, a wide variety of systems has been developed, primarily for land use planning and natural resource management at the urban, regional, state and national levels of government, but also for applications by public utilities and private corporations. Most of these systems rely on data from existing maps or on data that can be readily processed to provide the locational information required (Shelton and Estes 1979). At present, however, we estimate that only about ten commercial firms are offering fully integrated GIS on the open market in the United States. Furthermore, few academic institutions offer courses in GIS. In 1962, *Schwendeman's Directory of College Geography of the United States* listed only one department as offering a course in GIS, while in 1984 only four departments offered courses (Monsebroten 1982, 1984). This lack of academic training has inhibited the development of the field.

1.2. *Basic components of a GIS*

The function of an information system is to improve a user's ability to make decisions in research, planning and management. An information system involves a chain of steps from the observation and collection of data through their analysis to their use in some decision making process (Calkins and Tomlinson 1977). In this context, a GIS may be viewed as a major sub-system of an information system. A computer-based GIS may itself be viewed as having five component sub-systems (Knapp 1978), including

- (*a*) data encoding and input processing;
- (*b*) data management;
- (*c*) data retrieval;
- (*d*) data manipulation and analysis;
- (*e*) data display.

The five components are described in greater detail below.

1.3. *Requirements of large-scale GIS*

On the basis of recent research concerning the design and implementation of large-scale GIS, one may infer several requirements that any GIS possessing the above five components should satisfy, as well as several principles of design and implementation that permit the satisfaction of such requirements.

Recent research (see, for example, Marble 1982, Calkins 1983, Peuquet 1984) suggests that the following general requirements should be satisfied in the design and implementation of most GIS:

- (*a*) an ability to handle large, multilayered, heterogeneous data bases of spatially-indexed data;
- (*b*) an ability to query such data bases about the existence, location and properties of a wide range of spatial objects;
- (*c*) an efficiency in handling such queries that permits the system to be interactive;
- (*d*) a flexibility in configuring the system that is sufficient to permit the system to be easily tailored to accommodate a variety of specific applications and users;
- (*e*) an ability of the system to 'learn' in a significant way about the spatial objects in its knowledge and data bases during use of the system.

Of the large number of GIS that have been constructed during the past 20 years, none has possessed the full generality of these requirements. The work of Smith, Peuquet, Menon and Agarwal (1986) is representative of recent approaches to constructing a system that satisfies these requirements to a significant degree.

1.4. *Principles for satisfying the requirements*

There are several general principles that may be applied in order to facilitate the design and implementation of a GIS satisfying the five requirements listed above.

- (*a*) A first principle, relating to all five requirements, involves the systematic application of techniques and approaches developed in a variety of sub-fields of computer science (CS). To date, few GIS have been designed and implemented on the basis of the systematic application of such knowledge. Sub-fields of CS appearing to have particular relevance for GIS include software engineering, data base theory, the study of algorithms and complexity, artificial intelligence, computer graphics and natural language processing.

(*b*) A second principle, relating to the first three requirements listed above, involves the integration of approaches and procedures developed in a variety of disciplines that are related to GIS. These disciplines include computer vision, image understanding, digital cartography and remote sensing.

(*c*) A third principle, relating to the third requirement, involves the application of procedures that reduce the search effort involved in answering queries, particularly by avoiding simple, exhaustive search strategies. As noted by Smith *et al.* (1986), responding to queries about complex spatial objects in a large data base is an inherently difficult computational task.

(*d*) A final principle, relating to the fourth requirement, is to construct GIS in such a way that they may be easily tailored to specific applications and/or users by the users themselves.

2. Traditional GIS

We may define a 'traditional' GIS as a system for the efficient input, storage, representation and retrieval of spatially-indexed data. Two distinct classes of traditional GIS may be distinguished. The first class uses map-based data (particularly in vector format) and finds applications in, for example, engineering, boundary analyses and thematic representations. The second class uses image-based data (particularly in raster format) and finds applications in image analysis and remote sensing. Much research to date had involved GIS based on map-type data.

One may categorize the majority of queries typically handled by GIS into two classes. The first class of query requests the locations of some class of spatial objects within a given spatial window, while the second class of query requests the identities of objects found within a given spatial window and belonging to some sub-class of objects. Simple examples of these two fundamental queries include

(*a*) finding the location of sites for hazardous waste existing within some area;

(*b*) finding the geological objects that exist at some location.

It is, of course, possible to consider requests involving statistical summaries or analyses of the responses to these two major classes of queries, and to consider queries concerning changes over time of spatial objects in the data base.

In responding to queries, traditional GIS have typically involved the five components listed in § 1.2. We now summarize the main functions of these components.

2.1. *Data encoding and input processing*

Kennedy and Guinn (1975) describe automated spatial information systems as essentially data driven:

> while models which use the data are important to support the decision-making activities of those who use the system, a large portion of the investment will be obtaining, converting, and storing new data.

Data for input to a GIS are typically acquired in a variety of formats, including graphic data, non-spatial information (i.e. nominal data, descriptive or attribute data and textual data) from both printed and digital files, and digital spatial data tapes such as LANDSAT or digital elevation data. Often these data will require manual or automated pre-processing prior to encoding. During data acquisition, relevant information for each data type should be obtained which, in so far as possible, describes the accuracy, precision, currency and spatial characteristics (for example, geo-referencing system, scale) of the data (Kennedy and Meyers 1977, Salmen *et al.* 1977).

There is a variety of pre-processing procedures, including

- (*a*) format conversion;
- (*b*) reconstruction and generalization of data;
- (*c*) error detection and editing;
- (*d*) merging of points into lines and lines into polygons;
- (*e*) edge matching;
- (*f*) rectification and registration.

Format conversion covers many different problems, which can be classified into two different categories. The first category involves conversion between different data structures and the second involves conversions between different data media.

Within the context of a GIS, each spatial data type or theme is referred to as a spatial data layer or data plane. Within each of these spatial data layers there are four possible types of geographic entities which must be encoded: points, lines, surfaces and area-enclosing lines or polygons (Peucker and Chrisman 1975). For each spatial data layer, spatial objects may be encoded (or input directly in digital format) employing one of two basic techniques. In the first technique, entities may be generalized and stored using grid cells as a location identifier in a raster data structure. The generalization, however, often results in some loss of geographic specificity. If precise locational information is important, a second technique may be used in which entities are encoded and stored using any of several vector formats.

2.2. *Data management*

Data management allows a data base to be used through a combination of hardware and software facilities and operations. A GIS should include integrated data base management software designed to provide

- (*a*) the ability of the system to support multiple users and multiple data bases;
- (*b*) efficient data storage, retrieval, and update;
- (*c*) non-redundancy of data;
- (*d*) data independence, security and integrity.

Such systems are typically referred to as data base management systems (DBMS).

2.3. *Data retrieval*

In the creation of a data base, access procedures should be established to provide for retrieval of both spatial (graphic or image) and non-spatial (textual or attribute) data. Searches are conducted to locate features or entities or sets of features or entities. A GIS may be required to locate any of the following: a single feature; a set of defined features; an undefined feature or set of features (browse); features based on defined relationships within the data set; a set of features where the criteria are within another data set; and all features in a given class (Calkins and Tomlinson 1977, Salmen *et al.* 1977). Efficient data retrieval operations are largely dependent upon the volume of data stored, the method of data encoding and file structure design. In general, there are well-developed procedures for efficient retrieval of non-spatial data; however, searching for spatial features or sets of features often involves procedures of considerable complexity.

2.4. *Data manipulation and analysis*

Analysis of data from multiple spatial data planes typically requires the use of processing techniques capable of the manipulation and analysis of both grid and x, y-coordinate structured data and for conversion from one structure to another (Cicone 1977, Peuquet 1977, Marble and Peuquet 1977). Capabilities should also exist for the manipulation and analysis of attribute and tabular data files. The following represents a very brief discussion of a number of GIS manipulative and analytic procedures.

Data manipulation operations typically needed by users, and found in many GIS, include

  (*a*) reclassification and aggregation of attribute data;
  (*b*) geometrical operations such as rotation, translation and scaling of coordinates, conversion of geographic coordinates to specified map projections, rectification, registration and removal of distortion;
  (*c*) centroid and line allocation;
  (*d*) conversion of data structures;
  (*e*) spatial analysis of such properties as connectivity and neighbourhood statistics;
  (*f*) measurements of distance and direction;
  (*g*) statistical analyses.

Knapp (1978) groups procedures for data analysis into the following categories:

  (*a*) Spatial analysis, including procedures such as polygon overlay; cell overlay (arithmetic, weighted average, comparison, multiple map dependent reassignment, correlation functions); connectivity (proximity functions, optimum route selection, intervisibility); and neighbourhood statistics (slope, aspects, profile, clustering);
  (*b*) Measurement of line and arc lengths; of point-to-point distances; of perimeters, areas and volumes; of polygons in grid and x, y-coordinate format;
  (*c*) Statistical analysis, including histograms or frequency counts; regression, correlations and cross-tabulation; file generation for interface with a standard statistical package (such as SPSS, BIOMED, MINITAB);
  (*d*) Report generation, including the ability to provide labels for reports; to save text files as part of the data base; and to alter the standard default format.

Techniques of manipulation and analysis such as these operate on data retrieved from the spatial data base. New files containing 'value added' or derived data/information may be created for inclusion within the data base. These data can then be analysed at a later date.

2.5. *Data display*

A GIS should include software for the display of maps, graphs and tabular information in a variety of output media. Software should exist for the production of maps which depict the spatial or areal distribution of various objects and/or phenomena. The choice of which type of mapping to use depends upon a number of factors relating to the nature of the data (for example, representations of discontinuous or continuous distributions of features and/or phenomena) and the use to which the map will be put.

Display media include both hard copy materials (for example, paper products of various types, photographic products, or mylar) and hardware devices for the production of either temporary or permanent graphics. Hardware devices for output displays include line printers, ink line plotters, electrostatic printer/plotters, ink jet plotters, cathode ray tubes, colour film recorders and computer output microfilm devices (Knapp 1978).

The choice of GIS output hardware and software depends in part upon applications, of which there is a wide variety.

2.6. *Applications of geographic information systems*

The use of a GIS to satisfy the needs of a given user is called an application. Most GIS applications involve some form of geographic or spatial analysis. Basically we have moved from the applications of GIS technology for simple map overlay and comparison to complex spatial analyses (Estes 1984). GIS systems are now beginning to be employed to trace containment movement through the environment, to predict crop yields, to follow financial flows and to automate mapping and facilities management. This last area, for example, brings the civil engineering and architectural communities to GIS for their specific analytic needs. Yet all these applications involve, to one degree or another, mapping, monitoring and modelling activities.

Currently, most users involved in geographic analyses require maps as output. Along with significant federal and state requirements for thematic products, the business community has definite requirements for thematic mapping. In particular, private concerns which supply information derived from remotely-sensed data have found that almost all clients' requirements are, to a large extent, cartographic in nature (Simonett 1976).

Map accuracy is still, however, an important and unresolved research issue, and assessment of the accuracy of thematic maps is particularly difficult. Several publications deal with the subject of assessing the accuracy of a given thematic map (Rosenfeld *et al.* 1981, Rosenfeld 1982, Estes *et al.* 1984). The problem is compounded in GIS applications where thematic or other maps with differing accuracy attributes are combined to produce a new output. In addition, because monitoring inherently involves detection of change, and because modelling can involve analyses of processes which add their own variance to data, a user can readily appreciate the importance of knowing the accuracy associated with various GIS data inputs. Basic research, then, is needed to test the sensitivity of models to errors of both labelling and positioning in input data sets. How these errors propagate and affect the accuracy of output can have serious application-dependent consequences.

3. Technical issues in the design and construction of large-scale GIS

In the preceding sections we have listed the structure and functions of traditional GIS. Larger volumes of data and an increasing potential for use in diverse areas make it necessary to focus on improving the capacity and efficiency of such systems. We noted above that there are several generally-applicable principles for the design and implementation of efficient, large-scale GIS. An important subset of principles involves the systematic use of techniques developed in certain sub-fields of computer science. We now discuss in greater detail the applicability of these sub-fields to the design and implementation of GIS, as well as considering several other important issues.

3.1. *Software engineering*

Software engineering provides a set of techniques to aid in the design, implementation and testing of large software systems. Only recently have GIS researchers (such as Aronson 1984, Calkins 1983 and Marble 1983) described the applicability of techniques of software engineering to the construction of GIS.

Among the contributions of such techniques for the design, implementation and testing of large-scale GIS has been a set of models of the software 'lifecycle' (see, for example, Boehm 1981). Such models indicate the series of defined project phases, activities and milestones as well as the tools to be selected and applied at various phases of the project. Software engineering is also able to provide models that enable one to estimate the cost of producing a given software system. Marble (1982) has indicated that many of the unfortunate characteristics of past GIS development, including high failure rates, inflexibility, inability to adapt to different operating environments and problems of maintenance, could be avoided with the appropriate application of methodologies and tools of software engineering. As noted by Marble (1983), however, it may be important to specialize the tools developed for general software systems to the requirements of developing GIS. Little research has been performed in this area.

3.2. *Spatial data models and data structures*

The ability to take the spatial location of objects into account during search, retrieval, manipulation and analysis lies at the core of a GIS. The degree of success with which these tasks can be accomplished is determined by the data models, data structures and algorithms and data base management techniques selected for the GIS. The theory of spatial data bases is thus, in our opinion, the single most important area of CS that has relevance to the construction of GIS, and we have accordingly devoted a relatively large amount of space to this topic.

The characteristic feature of geographic data is that they are spatially indexed. Two basic alternatives are available in the construction of a data model that incorporates the spatial addressing:

(*a*) objects may be represented with each object having spatial location as an essential property;

(*b*) locations may be represented with each location being characterized by a set of object properties.

These alternatives may be seen as complements of each other, and have resulted in the vector and tessellation models respectively. Recent developments have also led to hybrid models (Peuquet 1983) which attempt to combine both, often within a relational framework that has traditionally been used in non-spatial data base applications.

3.2.1. *Vector models*

The basic logical unit in a vector model is the line, used to encode the locational description of an object, and represented as a string of coordinates of points along the line. Closed areas, modelled as polygons, are represented by the set of lines that constitute their boundaries. Vector models of geographic space may be classified as unlinked or topological.

3.2.1.1. *Unlinked model.*
The simplest form of vector representation has been termed the spaghetti model. In this model each map entity is encoded separately in vector form without referencing any of its neighbouring entities. Spatial relationships are not encoded, rendering spatial analysis of data expensive. Such models are, however,

adequate for display, and hence find application in computer cartography and graphics.

An alternative to representing points along the line is the use of the chain code (Freeman 1974). The location of a starting point on the boundary is represented and successive points are stored in terms of displacements from the starting point. Run length encoding can be used to achieve significant data compression. Geometric transformations and spatial relations are difficult to compute using vector data in chain code format.

3.2.1.2. *Topological model.* In the topological model the network of lines partitioning a map is represented as a planar graph. Line segments correspond to arcs in the graph and their endpoints to nodes. The representation chosen encodes topological information on the structure of the graph.

An example of a data structure organized on these principles is that used in the Dual Independent Map Encoding (DIME) geographic base files of the U.S. Bureau of the Census. The records in the DIME files represent segments with associated directions. For each segment, the 'from' and 'to' nodes are stored, as is a reference to the regions on the left- and right-hand sides of the segment when travelling in the assigned direction. The topological information stored makes it possible to automate procedures for checking consistency that detect errors which may be introduced during encoding. Such encoding also enables efficient analyses to be performed on the network. Optimal configurations of networks, including shortest paths and maximum commodity flows, are naturally studied using such representations.

A hierarchical representation based on the topological model is the POLYVRT data structure used to represent the partition of an area into polygons (Peucker and Chrisman 1975). The basic data entity in this scheme is the chain consisting of straight line boundary segments that have common regions on their left and right sides. A set of chains forms the boundary of a region, and hierarchical relationships between regions are used to organize the data. This procedure permits rapid access to the polygonal definitions of each area and allows selective retrieval in addition to revealing spatial relationships between regions.

3.2.2. *Tessellation models*

A tessellation of the plane is an aggregate of cells that partition the plane. The logical unit of data in tessellation models is thus a unit of space. Each unit of space has an associated set of object properties. The essential property of such a data model is that spatial relations between logical units are implicit in the tessellation. Data storage can thus be made to mirror arrangements in geographical space. The location properties of polygonal objects are not directly available and must be assembled by aggregating spatial units.

Planar tessellations that are useful for representing spatial data should satisfy two criteria:

(a) the tessellation should result in an infinitely repetitive pattern in the plane;
(b) the tessellation should be infinitely recursively decomposable into similar patterns of smaller size.

The first property allows data bases of any size to be represented, while the second makes it possible to use hierarchical data structures. The three regular tessellations of the plane, with squares, triangles and hexagons as the smallest logical units, all satisfy the first property. In any of these tessellations all cells have the same number of sides and the same number of cells meet at any vertex. Eight other semi-regular tessellations exist that satisfy the first property, but which are not made up of identical regular polygons (Ahuja 1983). Only the square and triangular tessellations are infinitely decomposable into similar patterns. The hexagonal tessellation offers the advantage of uniform adjacency and uniform orientation, making it useful for modelling applications involving radial distance propagation.

The most widely used tessellation is based on the square and corresponds to the raster data structure. The raster is easily traversed in both the x and the y directions in the uncompacted state. Run length encoding in the x direction is frequently used to achieve data compression. Most scanning devices used for data capture, such as multispectral scanners on satellites and optical scanners used for the mass digitizing of analogue maps, produce raster data sets.

3.2.2.1. *Hierarchical tessellation models.* A hierarchical tessellation model is based on a regular recursive decomposition of space. The square tessellation is most commonly used since the cells in the triangular tessellation do not have the property of uniform orientation. Since hexagons cannot be decomposed into identical hexagons only a partial hierarchical decomposition can be based on the hexagonal tessellation (Gibson and Lucas 1982).

(*a*) *Regional representation.* The recursive decomposition of the square cell gives rise to a tree with a branching factor of four. Such a tree offers a hierarchical view of the data that, depending on the actual data structure used in implementation, offers several potential advantages:

   (i) spatial relations remain implicitly coded in the hierarchical representation;
   (ii) efficient spatial search for desired object attributes can be carried out, since the higher-level views of the data allow efficient pruning in space;
   (iii) data compaction may result when four sibling cells at any one level in the tree sharing the same attribute value are omitted and are replaced by their parent;
   (iv) the higher levels in the tree may be used to browse through a given data set;
   (v) the structure of the decomposition admits efficient storage of, and access to, the data by areas.

A number of different data structures can be used with the hierarchical square tessellation model and they can be differentiated on the basis of the type of data represented, the principle guiding the decomposition, and the resolution characteristics of the resulting structure. The two principal hierarchical data structures in use are the quadtree and the pyramid. The quadtree is a variable resolution data structure, where the leaves of the tree may be at any level depending on the level of sibling aggregation taking place. The pyramid is a complete structure and stores all nodes in the hierarchy. Higher nodes in the pyramid store a data value based on averaging or some other function of the values of the descendants. The pyramid is thus a multiple resolution structure.

Initial work concentrated on pointer-based implementations for binary images (Samet 1980 a, 1980 b). The introduction of an addressing scheme replacing a leaf by a location code encoding the path taken to reach it from the root node enabled the introduction of a space-efficient version of the quadtree, termed a linear quadtree, in which only leaves were stored (Gargantini 1982). Hierarchical navigation and spatial operations are still possible because of the nature of the addressing scheme introduced.

Algorithms for set operations, such as intersection and union, connected component labelling and geometrical transformations on both forms of quadtree, are available. Algorithms to run–encode linear quadtree data have also been developed, resulting in further data compaction (Mark and Lauzon 1984).

In integrated GIS that handle both map and image data, a hybrid structure based on the virtues of both the quadtree and the pyramid offers advantages (Jackson 1986, Smith *et al.* 1986).

Large spatial data bases which are based on a hierarchical data structure require efficient means of storage of the data base on secondary storage and the capability of rapidly accessing relevant portions of the data base. The solutions adopted usually involve the use of a disk file structure (such as the B-tree) that allows only the data under scrutiny to be read into the main memory (Samet *et al.* 1984). The spatial addressing schemes outlined above allow efficient indexing schemes to be built for such retrieval.

(*b*) *Point and line representation.* Point data have been held in tree data structures based on both data-dependent and regular recursive decompositions of space. These include the k-d tree (Bentley 1975) and a number of versions of the quadtree. For a complete survey the reader is referred to Samet (1984).

Quadtrees based on the regular decomposition of space have been used for linear features. These models assume that the linear feature is embedded in a grid. The edge quadtree (Shneier 1981) subdivides the blocks of a regular quadtree till a square is obtained that contains a single curve that can be approximated by a straight line.

Strip trees (Ballard 1981) are a data-dependent, unlinked representation in which a single curved line is approximated by a series of rectangles, spatial relations between curves not being stored. The strip tree has found applications in computer vision and robotics, and is useful for queries involving search and set operations such as intersection.

In applications of GIS there is always a need to perform operations that require analysis of both line and region data. An example is the retrieval of all areas that grow corn within a particular distance of a river. Within the framework of hierarchical tessellation models it is advantageous to have regular decompositions for linear features if such queries are to be efficiently answered, as these are 'registered' to the representation for region data. This is the principal advantage of data structures like the edge quadtree over the strip tree for geographical applications.

3.3. *The relational data model*
The relational data model has been extensively used in applications in commercial data base management. Such systems offer the advantages of tabular structure, a set-oriented user interface, the ability to access data based on values and a high level of data independence. A relation is a subset of the cartesian product of n domains, and each element of the relation is stored in an n-column table. The relational data base can be queried in a structured manner using queries framed in the relational calculus. As an example, a particular n-tuple (row) in a relation can be selected based on the values of any particular column, and projected to obtain the values of a specific attribute.

The structure and simplicity of the relational model, including the complete absence of pointers, has also made it an attractive choice for GIS (van Roessel and Fosnight 1984). The topological vector models discussed earlier naturally lend themselves to use in relational data bases. Points, segments and polygons may be modelled as relations, with each entity being given a unique identifier. This allows attribute information also to be expressed using relations based on the same identifiers. An example of the application of such principles is the ARC-INFO system (ESRI 1984, Dangermond 1983). Such representations also allow for automatic procedures for maintaining consistency to be constructed (Meir and Ilg 1986).

In a GIS, geometric and topological relations exist between the different geographic entities, be they points, lines or regions. There are so many potential relations that it is not possible to store them entirely explicitly within the system. Relations may be stored implicitly using coordinate information, but retrieval then becomes more expensive. Extensions to relational query languages have been derived to handle queries such as the retrieval of all map elements within a window (Frank 1982).

Other generalized spatial data structures based on the relational model have also been proposed (Shapiro and Haralick 1980). A logical data structure is used to represent a geographic entity independent of the representation chosen for its location, whether vector- or raster-based.

3.4. *Algorithmic considerations*

The study of algorithms and complexity is applicable to GIS in its provision of a theoretical basis for algorithms that search large spatial data bases for complex spatial objects in an efficient manner. In particular, the emerging sub-field of computational geometry (see Preparata and Shamos 1985) promises much in the way of efficient spatial algorithms. The complexity of search problems encountered in geographical applications, together with the large size of the data bases used, warrants close study of the algorithms used within a GIS. The complexity of an algorithm may be measured in terms of the time taken, expressed as a functions of the 'size' of the input to the problem. In most cases faster query times can be obtained by judiciously pre-processing the data and storing them in an appropriate data structure. In these cases the storage consumed by the data structure and the time spent for pre-processing must also be taken into account. Most spatial queries can be solved with varying speeds depending on the level of pre-processing that is used.

Each of the data models discussed is appropriate in different situations. A comparison of algorithms for certain selective tasks is presented below.

3.4.1. *Overlay*

The overlay of two or more coverages over a selected area is a fundamental operation that must be performed in a GIS. The tessellation models offer conceptually simple algorithms, since the basic logical unit is a unit of space. Algorithms to overlay vector-based coverages are more complex. Raster overlays may still be time-consuming because of the volumes of data involved. Hierarchical data structures and two-dimensional run encoding can offer significant advantages. The result of any tessellation overlay still has a unit of space as the basic logical unit, and cells are not aggregated to form object entities. Complete object descriptions are returned from vector overlay on polygonal networks. The problem of (vector) polygon intersection has received considerable attention in computer graphics, especially in connection with algorithms for the removal of hidden lines and surfaces. With the proper choice of data structures and the use of sweep line techniques, efficient algorithms are possible (White 1978).

3.4.2. *Geometric searching*

The planar point location problem deals with the location of a point in a planar subdivision. The topological vector model is naturally adapted to such problems, the polygonal network being viewed as an *n*-vertex planar graph. Well-known methods include the chain method (Lee and Preparata 1977), the asymptotically optimal triangulation refinement method (Kirkpatrick 1983) and later methods, such as an optimal version of the chain method (Edelsbrunner *et al.* 1986).

The range search problem consists in retrieving all points from a given set that fall within some standard geometric shape arbitrarily translated in the coordinate space. In the general problem each coordinate could represent any non-spatial numerical key. The principal methods used to solve the problem when the query region is an orthogonal rectangle have been based on tree data structures, including the k-d tree (Bentley 1975) and the range tree (Willard 1978, Luecker 1978).

The rectangle cover problem, which consists of the retrieval of those rectangles in a given set that intersect a given query rectangle, is of relevance in the search for spatial entities within a sub-region. The problem has been addressed using a key based on a regular hierarchical decomposition (Abel and Smith 1983).

3.4.3. *Proximity analyses*

Several problems concerning point data sets have received intensive attention in computational geometry including the closest pair, all nearest neighbour, Euclidean minimum spanning tree, triangulation and nearest neighbour search problems. While the 'divide and conquer' paradigm has been used to attack the closest pair problem in multi-dimensional space, attacks on the remaining problems have been almost exclusively based on the construction of the Voronoi diagram (Preparata and Shamos 1985). The diagram is represented as a planar straight line graph and this lends itself to a network-based representation. Numerous generalizations of the diagram to handle lines, circles, further points and *k* nearest neighbours may be found in the literature.

The tessellation- or grid-based models offer advantages in several proximity analyses involving spatial constraint propagation because of their implicit spatial structure.

3.4.4. *Raster-based operations*

Several algorithms from image processing, segmentation and computer graphics are naturally suited to the tessellation data model, given the basic nature of the digital image. These include a variety of algorithms for spatial filtering, image classification, the creation of three-dimensional displays from combined elevation and image data, the derivation of slope and aspect images from terrain models, the creation of shaded relief models, automatic delineation of drainage basins and many other raster-based algorithms.

3.5. *Knowledge-based approaches*

Artificial intelligence (AI) studies computational techniques for solving problems which are either computationally intractable or for which there are no well-understood algorithms. The complexity of spatial objects and the size of the spatial data bases suggest the applicability of AI techniques in designing data structures for representing complex spatial objects; in developing procedures for reasoning about spatial objects and for answering various queries about the objects; and in designing 'control strategies' that determine when procedures should operate on given data structures in order to accomplish a given computation.

It is possible here to suggest only a few of the many implications that AI research has for the design and implementation of GIS. An example of a GIS which implements several of the data structures, procedures and control strategies developed by AI researchers is provided by Smith *et al.* (1986). Concerning the representation of spatial objects, there has been much research in AI on the use of languages based on predicate calculus for representing complex objects. One contribution of the research of Smith *et al.* has been the construction of such a language for the description of complex spatial objects. An advantage of this approach is the fact that the predicate calculus contains inference rules that may be applied in an 'automated' manner. AI has also led to the development of special-purpose data structures, such as semantic networks and frames, in which certain specialized forms of reasoning are efficiently implemented.

The classes of procedures developed by AI researchers that have applicability for GIS are too numerous to discuss systematically. In particular, however, we mention the development of efficient procedures for search, image processing and learning. Among the important search procedures are those that are guided by heuristic knowledge of the problem domain. The 'A*' procedure (see Nilsson 1980), for example, has found numerous applications in problems of planning spatial paths, and uses domain-specific knowledge to reduce average search times for finding an optimal solution. AI researchers have also investigated several systematic search procedures that have found application in GIS, including constraint satisfaction procedures (see Smith *et al.* 1986) and hierarchically-controlled procedures (see Smith and Parker 1986).

The contributions of AI researchers in the domain of image processing are of great relevance to GIS, and in particular to those based on raster data structures. As one example, we mention relaxation labelling of the objects in an image (see Ballard and Brown 1982) for several such procedures. The importance of these procedures will become of major significance as GIS are implemented on large, multi-processor systems. Finally we mention the learning procedures that have been developed by AI researchers. Smith *et al.* have employed a system originally developed by Hoff *et al.* (1983) that is able to learn new conceptual definitions from examples. In this application of learning in a GIS framework, the system is able to modify its knowledge base of spatial objects on the basis of experience.

There are several control structures that have been investigated by AI researchers that have relevance in a GIS context. The use of 'production systems', in which control is exercised by choosing the most applicable rule in a knowledge base of rules, is a natural method for modelling such diverse tasks as the search of large complex data bases of maps and images and the labelling of output images in an easily-interpretable manner.

In summarizing the applicability of AI techniques to the construction of future GIS, it is fairly safe to assume that they will come to play an increasingly important role, since GIS involve inherently difficult computational tasks.

3.6. *Integrated GIS*

An important principle relating to the design and implementation of modern GIS involves the integration of approaches and procedures developed in a variety of disciplines that are related to GIS. These disciplines include computer vision, image understanding and digital cartography (see, for example, Ballard and Brown 1982). Two reasons for this integration are as follows:

(*a*) These disciplines all study the same basic problem of recognizing and reasoning about spatial objects implicitly encoded in spatially-indexed data sets. Since their evolution has been somewhat independent, research on GIS would benefit from the integration of approaches and procedures developed in these other disciplines.

(*b*) There has been a recent and growing realization that it is often a practical necessity to merge image data sets, such as LANDSAT scenes, with the more traditional data sets of GIS, such as digitized maps and vectorized representations of map features (see Jackson 1985). Computer vision and image understanding have developed techniques that will allow the integration of such capabilities into GIS. Such systems must be able to integrate both raster- and vector-based approaches to spatial data analyses. Following Jackson, we may term such systems integrated GIS (IGIS).

The integration of map and image data within a single system is currently the focus of work at several research centres. The main thrust of current work in integration has involved the use of both hierarchical data structures and knowledge-based approaches to search and analysis.

3.7. *Input/output using graphic and natural language interfaces*

Computer graphics and natural language processing are sub-fields of computer science that provide techniques for constructing efficient and appropriate interfaces to a GIS. Early systems were user hostile in the sense that processing steps were not interactive, command procedures were abstruse and errors of job control syntax could cause an entire job to fail. Modern methods of system design and modern paradigms of man–machine interaction can produce efficient and robust systems that do not require great expertise to operate.

Most users of spatial data are familiar with a range of graphic representations before they are introduced to a GIS. Maps of many kinds and aircraft imagery are the fundamental tools of non-automated analysis of spatial data. In an automated system, these same graphics should be available to the user. Unlike early systems where graphics displays were not available, modern systems often have capabilities for vector or raster computer graphics for both soft and hard copy formats. These devices are needed during data input to validate entry of the data, during data analysis to monitor the process and generate new hypotheses, and for communicating the results of the processing to others.

Recent developments in natural language processing have direct relevance to the user interface for a GIS. Frequently, a GIS will be used by a diverse user community. Some of these users may not have a background in cognate disciplines (such as geography, computer graphics, cartography or geometry). Others may use the system only infrequently. An interface which permits users to interact with the system in natural language can dramatically decrease the time it takes to learn to operate the system, particularly for non-expert users. It is now possible for a GIS designer to choose from a large number of 'off-the-shelf' natural language interfaces.

3.8. *Hardware considerations*

Conventional computer hardware has undergone great changes in the past few years. These have dramatically affected the cost of data processing. Peterson (1984) estimates that over the past 20 years, the cost of scientific computation has decreased by an order of magnitude every 8 years. This trend has accelerated recently, in large part owing to new, powerful microprocessors. Recent introductions of workstations suggest a price of $13 000 per MIPS (Collett 1986). As these trends accelerate, the complex computations in a GIS become cost effective for many new kinds of users.

There are specialized hardware systems in current practice that are not frequently applied to GIS. Single instruction–multiple data array processors have found limited application in commercial GIS. When array processors are used on a regular basis, they can provide a great increase in computational ability for modest cost. Their limited application is surprising, particularly when these devices are such a natural complement to raster data structures.

Specialized array processors are sometimes used in image processing. Adams *et al.* (1984) discuss the design and performance of this kind of hardware for geometric processing of raster data. In the example that they present, dedicated hardware performs 16 times faster than the conventional super-minicomputer.

Future hardware architectures may permit improvements in the environment for GIS in several fundamental ways. New workstations are being developed specifically for networked problem solving. In this environment, heterogeneous networks will bring the data from central repositories to the user's workstation for local analysis and manipulation. Expensive and specialized I/O devices such as scanning digitizers and colour filmwriters can be shared by many users in such a structured environment. Specialized hardware, such as parallel and pipelined processors, can provide dramatically improved performance for such tasks as geometrical operations, conversion of data structures and searches of large data bases.

Finally, new computer architectures involving massive networks of interconnected processors have applications to GIS. These relate in part to the problems of machine vision discussed earlier. In addition, it should be possible to analyse such problems as network flows on these massively parallel networks, in which each node in the network has its own dedicated processor. In such circumstances, the system may approach real-time performance, lowering costs to the point where many problems become tractable.

4. Conclusions

There are two trends relating to the use of GIS that appear self-evident:

(a) the rate of generation of spatially-indexed data, both from a large variety of sources and concerning a broad array of geographical phenomena, will continue to increase over the next few decades, leading to huge volumes of data for storage, retrieval and analysis;

(b) the demand for GIS to handle such volumes of data in a large variety of decision-making situations will also increase dramatically.

Given these trends, it is important that the design, implementation and use of GIS be placed on a more systematic and scientific basis than has generally been the case until recently. Such a basis, we believe, involves the application of the theory and techniques of several sub-fields of computer science and the integration of techniques developed in computer vision and image processing in the design and implementation of GIS. In particular, we believe that a great deal of investigation remains to be done concerning appropriate data structures and computational procedures for the storage, retrieval and analysis of spatially-referenced data in large-scale GIS.

Acknowledgment

This work was partially supported by a grant from NASA under NAGW-455.

References

ABEL, D. J., and SMITH, J. L., 1983, A data structure and algorithm based on a linear key for rectangle retrieval. *Computer Vision, Graphics and Image Processing*, **24**, 1.

ADAMS, J., PATTON, C., READER, C., and ZAMORA, D., 1984, Hardware for geometric warping. *Electronic Imaging*, **3**, No. 4, 50.

AHUKA, N., 1983, On approaches to polygonal decomposition for hierarchical image representation. *Computer Vision, Graphics and Image Processing*, **24**, 200.

ARONSON, P., 1984, Applying software engineering to a general purpose GIS. *Proceedings of Auto-Carto 7* (Washington: American Society of Photogrammetry), p. 23.

BALLARD, D. H., 1981, Strip trees: A hierarchical representation for curves. *Communications of the ACM*, **24**, 310.

BALLARD, D. H., and BROWN, C. M., 1982, *Computer Vision* (Englewood Cliffs, New Jersey: Prentice Hall).

BENTLEY, J. L., 1975, Multidimensional binary search trees used for associative searching. *Communications of the ACM*, **18**, 509.

BOEHM, B. W., 1981, *Software Engineering Economics* (Englewood Cliffs, New Jersey: Prentice Hall).

CALKINS, H. W., 1983, A pragmatic approach to GIS design. International Geographic Union Commission on Geographical Data Sensing and Processing, New York.

CALKINS, H. W., and TOMLINSON, R. F., 1977, Geographic information systems, methods and equipment for land use planning. International Geographic Union Commission on Geographical Data Sensing and Processing, Resource and Land Investigations (RALI) Program, U.S. Geological Survey, Reston, Virginia.

CICONE, R. C., 1977, Remote sensing and geographically based information systems. *Proceedings of the 11th International Symposium on Remote Sensing of Environment*, **2**, 1130.

COLLETT, R. E., 1986, MIPS battle heats up in workstation segment. *Digital Design*, **16**, No. 9, 19.

DANGERMOND, J., 1983, A classification of software compoments commonly used in geographic information systems. In *Design and Implementation of Computer-Based Geographic Information Systems*, edited by D. Peuquet and J. O'Callaghan (New York: Amherst).

EDELSBRUNNER, E., GUIBAS, L. J., and STOLFI, J., 1986, Optimal point location in a monotone subdivision. *Siam Journal of Computing*, **15**, 317.

ESRI, 1984, ARC/INFO: A modern geographic information system. In *Basic Readings in Geographic Information Systems*, edited by D. F. Marble, H. W. Calkins and D. J. Peuquet (Williamsville, New York: SPAD Systems Ltd).

ESTES, J. E., 1984, Improved information systems: a critical need. *Proceedings of the 10th International Symposium on Machine Processing of Remotely Sensed Data*, Laboratory for Applications of Remote Sensing, Purdue University, p. 2.

ESTES, J. E., SCEPAN, J., RITTER, L., and BORELLA, H. M., 1984, Evaluation of low-altitude remote sensing techniques for obtaining site information. NUREG/CR-3583, S-762-RE, Nuclear Regulatory Commission, Washington, D.C.

FRANK, A., 1982, MapQuery: Data base query language for retrieval of geometric data and their graphical representation. *Computer Graphics*, **16**, 199.

FREEMAN, H., 1974, Computer processing of line drawing images. *Computing Surveys*, **6**, 57.

GARGANTINI, I., 1982, An effective way to represent quadtrees. *Communications of the ACM*, **25**, 905.

GIBSON, L., and LUCAS, D., 1982, Vectorization of raster images using hierarchical methods. *Computer Graphics and Image Processing*, **20**, 82.

HOFF, W., MICHALSKI, R. S., and STEPP, R., 1983, INDUCE/2: A program for learning structural descriptions from examples. UIUCDCS-F-83-904, Department of Computer Science, University of Illinois at Urbana Champaign.

JACKSON, M. J., 1985, The development of integrated geo-information systems. Paper presented at Survey and Mapping 1985, Reading, England.

JACKSON, M. J., 1986, The development of integrated geo-information systems. *International Journal of Remote Sensing*, **7**, 723.

KIRKPATRICK, D. G., 1983, Optimal search in planar subdivisions. *Siam Journal of Computing*, **12**, 28.

KENNEDY, M., and GUINN, C., 1975, Automated spatial data information systems: Avoiding failure. Urban Studies Center, Louisville, p. 76.

KENNEDY, M., and MEYERS, C. R., 1977, Spatial information systems: Introduction. Urban Studies Center, Louisville.

KNAPP, E., 1978, Landsat and ancillary data inputs to an automated geographic information system. Report No. CSC/tr78/6019, Computer Science Corporation, Silver Springs, Maryland.

LEE, D. T., and PREPARATA, F. P., 1977, Location of a point in a planar subdivision. *Siam Journal of Computing*, **6**, 594.

LUECKER, G. S., 1978, A data structure for orthogonal range queries. *Proceedings of the 19th Annual Symposium on the Foundations of Computer Science*, p. 28.

McHARG, I., 1969, *Design With Nature* (Garden City, New Jersey: Doubleday).

MARBLE, D. F., 1983, On the application of software engineering technology to the development of geographic information systems. In *Design and Implementation of Computer Based Geographic Information Systems*, edited by D. Peuquet and J. O'Callaghan (New York: Amherst).

MARBLE, D. F., and PEUQUET, D. J., 1977, Computer software for spatial data handling: Current status and future development needs. Geographic Information Systems Laboratory, State University of New York, Buffalo.

MARK, D. M., and LAUZON, J. P., 1984, Linear quadtrees for geographic information systems. *Proceedings of the International Symposium on Spatial Data Handling*, Zurich, edited by D. F. Marble, K. E. Brassel, D. J. Peuquet and H. Kishimito (Geographisches Institut Abteilung Kartographie/EDV, Universität Zurich-Irchel).

MEIR, A., and ILG, H., 1986, Consistent operations on a spatial data structure. *I.E.E.E. Transactions on Pattern Analysis and Machine Intelligence*, **8**, 532.

MONSEBROTEN, D. R. (editor), 1982, *Schwendeman's Directory of College Geography of the United States*, Vol. 33, No. 1, Geographical Studies and Research Center, Department of Geography, Eastern Kentucky University, Richmond, Kentucky, p. 84.

MONSEBROTEN, D. R. (editor), 1984, *Schwendeman's Directory of College Geography of the United States*, Vol. 35, No. 1, Geographical Studies and Research Center, Department of Geography, Eastern Kentucky University, Richmond, Kentucky, p. 92.

NILSSON, N. J., 1980, *Artificial Intelligence* (Palo Alto, California: Tioga Press).

PETERSON, V. L., 1984, Impact of computers on aerodynamics research and development. *Proceedings of the Institute of Electrical and Electronic Engineers*, Vol. 72, No. 1, p. 68.

PEUCKER, T. K., and CHRISMAN, N., 1975, Cartographic data structures. *The American Cartographer*, **2**, 55.

PEUQUET, D. J., 1977, Raster data handling in geographic information systems. SUNY Buffalo, Geographic Information Systems Laboratory, State University of New York, Buffalo.

PEUQUET, D. J., 1983, A hybrid structure for the storage and manipulation of very large spatial data sets. *Computer Vision, Graphics and Image Processing*, **24**, 14.

PEUQUET, D. J., 1984, A conceptual framework and comparison of spatial data models. *Cartographica*, **21**, 66.

PREPARATA, F. P., and SHAMOS, M. I., 1985, *Computational Geometry* (New York: Springer-Verlag).

ROSENFELD, G. H., 1982, Sample design for estimating change in land use and cover. *Photogrammetric Engineering and Remote Sensing*, **48**, 793.

ROSENFELD, G. H., FITZPATRICK-LINS, K., and LING, H. S., 1981, Sampling for thematic map accuracy testing. *Photogrammetric Engineering and Remote Sensing*, **48**, 131.

SALMEN, L., MUTTER, D. L., and BURNHAM, K., 1977, *A General Design Schema for an Operational Geographic Information System for the U.S. Fish and Wildlife Service Region Six* (Western Governors Policy Office, Fort Collins) (Washington, D.C.: U.S. Government Printing Office), p. 27.

SAMET, H., 1980 a, Region representation: Quadtrees from boundary codes. *Communications of the ACM*, **23**, 163.

SAMET, H., 1980 b, Region representation: Quadtrees from binary arrays. *Computer Graphics and Image Processing*, **13**, 88.

SAMET, H., 1984, The quadtree and related hierarchical data structures. *ACM Computing Surveys*, **16**, 187.

SAMET, H., ROSENFELD, A., SHAFFER, C. A., and WEBBER, R. E., 1984, A geographic information system based on quadtrees. *Proceedings of the International Symposium on Spatial Data Handling*, Zurich, edited by D. F. Marble, K. E. Brassel, D. J. Peuquet and H. Kishimoto (Geographisches Institut, Abteilung Kartographie/EDV, Universität Zürich–Irchel).

SHAPIRO, L. G., and HARALICK, R. M., 1980, A spatial data structure. *GeoProcessing*, **1**, 313.

SHELTON, R. L., and ESTES, J. E., 1979, Integration of remote sensing and geographic information systems. *Proceedings, 13th International Symposium on Remote Sensing of Environment* (Ann Arbor, Michigan: Environmental Research Institute of Michigan), p. 675.

SHNEIER, M., 1981, Two hierarchical linear feature representations: Edge pyramids and edge quadtrees. *Computer Graphics and Image Processing*, **17**, 211.

SIMONETT, D. S. (editor), 1976, Applications review for a space program imaging radar. Technical Report No. 1, NASA Contract No. NAS9-14816, Santa Barbara Remote Sensing Unit, Johnson Space Center, p. 6.

SMITH, T. R., and PARKER, R. E., 1986, An analysis of the efficiency and efficacy of hierarchical procedures for computing trajectories over complex surface. *European Journal of Operations Research* (in the press).

SMITH, T. R., PEUQUET, D. J., MENON, S., and AGARWAL, P., 1986, KBGIS-II: A knowledge based geographic information system. Technical Report TRCS 86-13, Department of Computer Science, University of California, Santa Barbara.

STAR, J. L., COSENTINO, M. J., and FORESMAN, T. W., 1984, Geographic information systems: Questions to ask before it's too late. *Proceedings of 10th International Symposium on Machine Processing of Remotely Sensed Data*, Purdue University Laboratory for Applications of Remote Sensing, p. 194.

TOMLINSON, R. F., 1982, Panel discussion: Technology alternatives and technology transfer. *Computer Assisted Cartography and Geographic Information Processing, Hope and Realism*, edited by D. H. Douglas and A. R. Boyle, Canadian Cartographic Association, Department of Geography, University of Ottowa, p. 65.

USFWS (U.S. DEPARTMENT OF THE INTERIOR, FISH AND WILDLIFE SERVICE), 1977, Comparison of selected operational capabilities of fifty-four geographic information systems. FWS/OBS 77/54, Biological Services Program.

VAN ROESSEL, J. W., and FOSNIGHT, E. A., 1984, A relational approach to vector data structure conversion. *Proceedings of the International Symposium on Spatial Data Handling*, Zurich, edited by D. F. Marble, K. E. Brassel, D. J. Peuquet and H. Kishimoto (Geographisches Institut, Abteilung Kartographie/EDV, Universität Zürich–Irchel).

WHITE, D., 1978, A design for polygon overlay. *First International Advanced Study Symposium on Topological Data Structures for Geographic Information Systems, Harvard Papers on GIS*, Vol. 6 (Spatial Algorithms: Efficiency in Theory and Practice), edited by G. Dutton, Laboratory for Computer Graphics and Spatial Analysis, Graduate School of Design, Harvard University.

WILLARD, D. E., 1978, Predicate oriented database search algorithms. Ph.D. thesis, Harvard University, Cambridge, Massachusetts.

SECTION 2
Data Capture

Overview

Acquisition and input of spatial data is typically the most costly and time consuming portion of GIS implementation. It is also a subject which is not very well covered in the literature. There is a need for more research and publications on both GIS data capture and spatial data models (structures) in GIS. In the following article, Dangermond reviews various options for acquiring digital spatial data for a GIS. Both the use of existing digital data and the digitizing of new data are covered.

Suggested Additional Reading

Burrough, P.A. 1986. Data input, verification, storage, and output. In Ch. 4 *Principles of Geographical Information Systems for Land Resources Assessment*. Oxford University Press. pp. 57–80.

Peuquet, D.J. 1984. A conceptual Framework and Comparison of Spatial Data Models. *Cartographica*. 21(4):66–113.

Jack Dangermond
Environmental Systems Research Institute
380 New York Street
Redlands, CA 92373

A REVIEW OF DIGITAL DATA COMMONLY AVAILABLE AND SOME OF THE PRACTICAL PROBLEMS OF ENTERING THEM INTO A GIS

ABSTRACT.   The paper first describes various means of acquiring digital data for GIS's, including an extended discussion of various existing digital data types which only require conversion. Then current methods for entering polygon, tabular, text and other data forms are discussed. The paper concludes with some ideas about the significance of technological developments in this area.

INTRODUCTION

For approximately 25 years, organizations have been creating digital data bases of spatial or cartographic information. With the rapid increase in software and hardware tools for managing spatial data, there has been a rather dramatic increase in information being loaded into the computer for spatial data processing. This phenomenon has involved many different fields of science, government and education, and has involved many applications ranging from automating the drafting function to using information for planning and management applications.

These digital data are often the most expensive part of the GIS's to which they belong. Yet, often, not enough attention is given to the quality of these data or the processes by which they are prepared for automation and then captured for the data base. While this was perhaps understandable when GIS's were mostly used on on individual, short term, isolated projects, the growing use of GIS's in support of long range efforts and the creation of multiple use, widely shared geographic data bases requires that the data and the data entry process be closely examined.

This paper reviews some of the data commonly used in these cartographic information systems, and some of the tools and techniques people are using to capture this information for creation of digital data bases, including the many different exchange formats and conversion procedures for information already in digital form.

Generally, this paper deals with GIS's having vector map information represented in topological data structures linked to attributes in a relational data base. ESRI's primary experience has been in using the ARC/INFO system, although our staff have also had experience using other vector GIS systems.

DIGITAL DATA FOR GIS'S

There are six basic kinds of systems which provide the digital information that are appropriate for interface to or integration into a GIS (see Figure 1):

TECHNOLOGY	GENERIC DATA STRUCTURE	SPECIFIC DATA FORMATS
COMPUTER AIDED DRAFTING SYSTEMS	GRAPHIC DATA (MAPS/DRAWINGS)	• COGO • SIF • CADD VENDORS (AUTOCADD, COMPUTER VISION...)
DIGITIZING AND PHOTO QUALITY SCANNING SYSTEM	SPATIAL FEATURE DATA (VECTORS)	• DLG • DIME/TIGER • ETAK • POINT MEASUREMENTS (X, Y, Zs) • MOSS/PIOS
IMAGE PROCESSING SYSTEMS	RASTOR DATA	• IMAGE DATA (REMOTE SENSING) • GRID CELL GIS • CELLULAR DTMS/DEMS
DBMS TABULAR FILE SYSTEMS	TABULAR DATA	• HIERARCHICAL • NETWORK • RELATIONAL/FLAT FILES
WORD PROCESSING SYSTEMS	TEXT	• BIBLIOGRAPHY DATA (INDICES) • DOCUMENTS • POSTSCRIPT PRIMITIVE
VIDEO/LASER SYSTEMS	IMAGE PICTURE	• VIDEO • DIGITAL

Figure 1

Computer aided drafting systems which capture and maintain maps as electronic drawings. These systems range anywhere from PC based systems to those based on minicomputers or mainframes.

Systems (such as those based on digitizing, scanning or photogrammetry) for capturing information and spatial features in vector format.

Image processing systems, which capture information in raster format.

Tabular data base management systems (DBMS's) and their related files.

Word processing systems for managing text.

Video and laser image systems which capture and manage pictures.

GIS's can be used to build cross-indexing relationships between and among these other data. This is done either through building relationships between geographic features and these other types of information or through integration of these data into a GIS data structure. In the former case the GIS simply indexes the information for rapid access by way of spatial coordinates (e.g., pictures along a road may be accessed from a video logging system by relation to geographic features on a digital map, or electronic drawings from a CAD system are accessed by touching a map feature on a GIS display). In the latter case the information is actually integrated into a vector GIS.

THE MODEL

The basic model for vector geographic information systems breaks down our perceptual and physical reality into three basic data types: 1) points, 2) lines, and 3) polygons. These three give the cartographic location of phenomena expressed typically in x,y coordinates associated with a geodetic grid or measurement system. Attached to these are attribute information describing the characteristics of these primitive locations.

Historically, map and geographic-related information has been organized and grouped in many ways. At ESRI we have found that the following six basic categories of information are useful for general topical discussion:

Basemaps containing geodetic control, topographic elevation and planimetric features.

Land ownership parcels and their attributes (land records).

Roads and other transportation (expressed as centerlines) and their attributes, such as paving, width, etc.

Administrative boundaries, such as census tracts, police districts, water districts, etc.

Natural resource information, such as soils, climatic information, hydrography, etc.

Engineering records, such as sewer lines, water lines, utility/facilities, etc.

CARTOGRAPHIC DATA ACQUISITION

There are several kinds of information (see Figure 2) which must be automated prior to inclusion in a GIS; these include manually drawn maps and manually measured data (such as surveys and legal records). Also, some cartographic information often already exists in automated form.

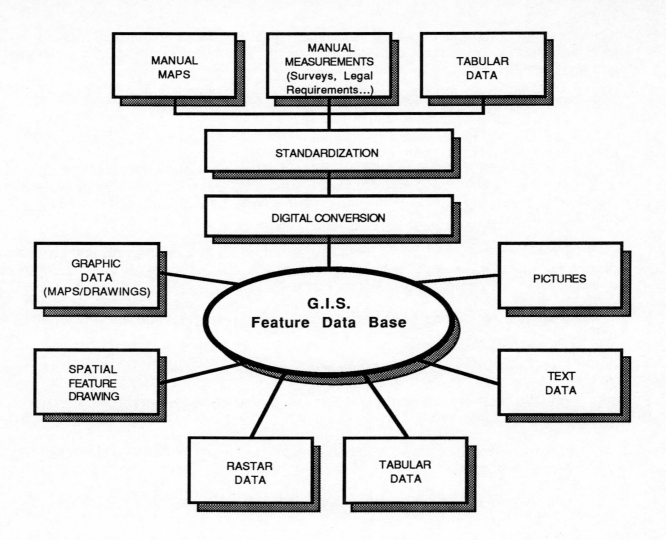

Figure 2

Thus there are at least four basic procedures for automating cartographic data. They are:

Manual digitizing

Automatic scanning

Entry of coordinates using coordinate geometry

Conversion from previously automated information.

Manual Digitizing

While considerable work has been done with newer technologies, the overwhelming majority of cartographic data entry is now done by manual digitizing. The reasons for this are many: one may not be able to remove the maps to where a scanner is available for doing the actual conversion; records may not be in a form that can be scanned (e.g., the maps are of poor quality, are in poor condition or have errors); the cartographic features may be too few on a single map to make it practical to scan; a scanner may be unable to distinguish the features to be captured from the surrounding graphic information on the display; scanning may not provide the required data precision (for certain applications, relatively high precision is required, and lower priced scanners simply do not offer the precision that is required); scanning may be more expensive than manual digitizing, considering all the cost/performance issues.

Manual digitizing has many advantages: low capital cost, low-cost labor and great flexibility and adaptability. While it is a time-consuming procedure, the technique can be taught to users within hours, and, with modern data base error checking software, the quality of the information is quite high. Interactive entry and editing can be done while users work on the cartographic data; errors on the basic map can be easily discovered and updated while in the process of entering the information; and digitizing devices are very reliable.

For these reasons, we see the majority of actual cartographic data entry still occurring by manual digitizing. While there is some hope for improvement, digital scanning will not fully replace manual digitizing for a considerable time.

Scanning

While scanners are expensive to acquire and to operate, scanning technology has been making major breakthroughs in its the ability to automatically capture information from maps. Our experience, however, has been that scanners work best when the information is kept very clean, very simple and uncluttered with graphic symbology. This is particularly true for entry into a GIS of data base primitives (i.e., points, lines and polygons), rather than just large volumes of graphics and text.

At ESRI we use scanners for all maps that we are able to re-draw from photointerpretation or from other maps and drawings. These are very clean manuscripts, not requiring post-processing for clean up of cartographic errors and other phantom graphic material that is captured by the scanner, but not required by the actual GIS.

Scanning is most appropriate for such maps, for maps which contain large volumes of cartographic feature information (for example, maps with 1,000 or more polygons) and maps whose cartographic feature definitions require substantial amounts of x,y coordinate definition (i.e., the lines are sinuous, such as irregular shaped soils, irregular shaped stream networks, etc.). Such maps are best scanned when there is only one feature type on the map (i.e., one parameter such as soils or vegetation).

Conversely, maps which are not clean and require interpretation or adjustment during the automation process or which have small numbers of cartographic features, are simply not worth scanning.

Coordinate Geometry

A third technique, particularly useful for entering land record information-- such as the entry of legal descriptions for property-- involves the calculation and entry of coordinates using coordinate geometry procedures (COGO). This requires entering, from survey data, the explicit measurement of features from some known monument; most often this includes the actual metes and bounds descriptions in distance and bearings, using alphanumeric instructions. This technique is useful for creating very precise cartographic definitions of property, and is particularly useful when the maps must represent exactly the land cadaster as it is expressed in the legal description. Surveyors and engineers like to use this technique because it provides them with very high levels of accuracy.

Unfortunately, the use of coordinate geometry is substantially more expensive than other means the entering cartographic data: anywhere from four to twenty times more expensive. Normally the cost is in the range of six times more expensive. A city with 100,000 parcels may spend something like $1.50 per parcel for digitizing or $150,000 total, but anywhere from five to ten times that amount if entering the information using coordinate geometry. This has created quite a controversy within local government between various users of potential GIS systems. The planners and, in fact, most of the users of the digital files, are quite willing to accept the level of accuracy provided by simple manual digitizing. But the engineering professions often want highly accurate coordinates for the land boundaries. When analysts have examined the benefits resulting from this increased accuracy, the results have been controversial. For the overall community of municipal users the benefits seem relatively small, particularly in the context of multi-user investment sharing over time; most local governments have taken the stand that the benefits simply do not outweigh the cost investments necessary. Engineers argue, however, that precision is necessary for survey and engineering computations.

Some confusion has arisen because many initial attempts at digital mapping were made with computer-aided drafting (CAD) technology, which aimed to create map "drawings" for engineering and drafting design. Often these technologies had been appropriately used on a "project" basis, but some CAD users advocated putting in very high precision right across a city or county without evaluating the resulting cost vs. multi-user benefits.

At present, the most popular scheme is the entry of some of the geodetic control information using COGO and other techniques, and then manual entry of actual subdivision lot boundaries or, in some cases, blocks. In Alberta, Canada, for example, COGO is being used to put in the x,y coordinates of block corners and block boundaries, while the actual lots are being manually entered by digitizing.

Many cities and counties have actually done their lot checking using COGO techniques. A few of the more enlightened organizations have saved these COGO definitions in digital form and have converted them into x,y coordinate form for use in a vector GIS.

COGO Updating

A popular idea in recent years is the manual digitizing of coordinates for an entire municipality and updating these coordinates with COGO over time, using transactions such as land subdivisions as the update mechanism. The procedure is analagous to the creation of an actual subdivision. Horizontal control is established first; then the area's boundary is tied into the overall network of surrounding parcels and finally the lot is subdivided using coordinate geometry tools. While attractive in theory, this does not work unless explicit control at a block or block grouping level has been established, allowing the updates to be COGOed in off of some monumentation nearby. Experience has shown that this approach requires close coordination among the departments responsible for the subdivision. The engineer and surveyor, tax assessor and recorder must be closely involved in the actual transactional updates to ensure that completeness, closure and high accuracy are retained.

ENTRY OF OTHER AUTOMATED FILES

There are over a dozen other standard types of digital cartographic files that can and are being used for building GIS systems. Each has various associated problems and opportunities which are described below.

DIME Files

The quality of DIME files in the United States is improving, particularly with the 1990 census. However, they are typically cartographically poor, and their address information is often inconsistent. They can be read directly into a GIS system and the address and related attribute information put into attribute files. Numerous digital tools, such as route tracing, can be used for address range consistency checking within the GIS in order to upgrade the quality of the basic DIME file. The DIME file coordinates are poor primarily because of the original scanning technique that was used by the Bureau of the Census in capturing the actual coordinates. While cartographically incorrect, they often have very high quality address ranges and, topologically, are mostly consistent. These data sets are very useful for address matching, address geocoding and, to a lesser extent, thematic mapping. They have the distinct advantage of being very inexpensive and, for very small funds, can be quite useful in very generalized urban analysis and mapping.

ETAK Road Information

ETAK files are a commercial data base of road centerline information used for road navigation by small processors inside of automobiles. ESRI has created GIS systems for cities and regions using this information. The information is typically based upon 1:24,000 map sheets, is cartographically quite accurate, topologically consistent and has very good quality address geocoding. Because it's been pre-cleaned with topological algorithms, it moves into ARC/INFO very quickly.

Digital Line Graph (DLG)

The DLG is produced by the USGS and reflects the cartographic information extracted from 1:100,000 and 1:24,000 map sheets for the United States. These maps are extremely good, particularly the 1:100,000 sheets, which now cover the entire United States. When the DLG's are overlaid against digitized 1:24,000 maps, the majority of data are either directly on or within a pixel of the 1:24,000's information. DLG information can be exchanged and directly pipelined into a GIS using DLG as the exchange format. DLG has the advantage of being topologically consistent and, in version 3.0, is capable of carrying considerable attribute information as well. Our experience with USGS DLG files has been that they are of very high quality, both in their cartography information and their attribute data.

Coordinate Geometry Data Bases

Local governments, particularly counties, have spent considerable time checking and verifying their parcel boundary information by use of various survey and coordinate geometry packages. A number of these organizations have retained the lot closure information in digital form and these data can be activated and converted into actual GIS polygon information through various software tools if there is good geodetic control referencing within the actual legal descriptions. This is usually so in the western states, where the public land survey coordinate township fabric is commonly used within the legal description. For a control network of the public land survey (township, section, quarter-section and quarter-quarter section) geographic coordinates may be necessary for initial entry using a coordinate geometry technique in order for the subdivision coordinate geometry descriptions to be able to be properly related to the overall fabric.

One of the problems with using historic COGO records is that only rarely have they been plotted out and verified so we have often found substantial underlaps and overlaps on the legal descriptions of properties. These have to be resolved and corrected, which can be a very time-consuming process. The approach for making these corrections can range from merely graphic "fudging" to resolution of the legally recorded inconsistencies, a much more expensive and complicated procedure.

Standard Interchange Format (SIF) for Various CAD Systems

Considerable investments have been made in digital mapping using various computer-aided drafting (CAD) tools. These systems generally use a standard interchange format (SIF) developed by the CAD technology manufacturers. This interchange format is primarily designed for moving CAD data structures between the various vendors, and adheres to the typical graphic primitives that are common among the interactive graphic systems (e.g., splines, circles, boxes, symbols, etc.). ESRI has had considerable experience in converting data (Figure 3) from the SIF format into the actual points, lines and polygons of a GIS . There are difficulties. Symbolization (e.g., shading of polygons or symbolization of points and lines) often has to be "stripped" off the cartographic data before topology building and data base creation. Also we have found that the graphic files have rarely been previously checked for topological consistency. They often contain line undershoots and overshoots, missing lines, gaps, etc., and have to be run through an automatic "cleaning" procedure in order to make them topologically consistent. While it is relatively simple to convert from a SIF file to a GIS file, it is somewhat more difficult to clean up the errors. One should anticipate some post-processing and editing. This, of course, is variable.

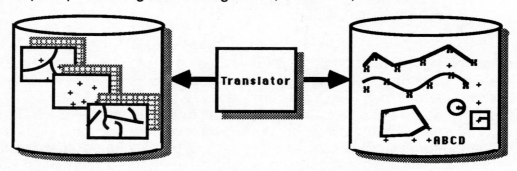

CADD INTERFACE

Figure 3

Polygon Based Systems

During the 70's, a number of GIS systems were developed which involved closing loops to define polygons-- MOSS and PIOS were two of these. The conversion of files from such systems is relatively straightforward, requiring computer time but almost no labor. To convert from these into modern topological data structures, users must redefine the polygons as a series of arcs and nodes. This is largely an automatic process. If there are a lot of splinters and sliver errors introduced by digitizing adjacent polygons, then, beyond certain tolerances, concepts of fuzziness won't work, and post-processing and interactive updating will be required.

Image/Cellular Files

ESRI brings land cover or other thematic layers obtained through image processed remote sensing into a vector GIS (see Figure 4) through a vectorization procedure known as GRID to ARC. In such a procedure, large groupings of homogenous pixels are delineated as polygons, each with attributes of area, perimeter and thematic codes. Problems encountered usually involve isolated pixels which have unique classifications; in some cases, these are simply remnants from the classification procedure and ought to be "eliminated". GIS procedures allow this to be done either as a pre- or post-processing effort. Other problems occur when these isolated pixels/cells define long sinuous phenomena like streams, or cliffs or other linear discontinuities. In these cases it has been necessary to apply smarter software which can recognize these sinuous phenomena and convert them into vector lines or bands representing the actual pixel phenomena.

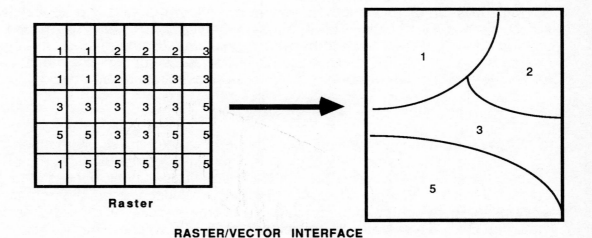

Raster

RASTER/VECTOR INTERFACE

Figure 4

Digital Elevation Models/Digital Terrain Models

There are at least four different ways to store digital elevation. They include contour lines with the attribute of elevation; random or equally-spaced x,y,z coordinates; a cellular digital terrain model; and a triangulated irregular network.

The first and second of these can be easily used in a vector GIS system; to use the third is more difficult. ESRI has developed software which selects "very important points" from a digital elevation model for formulation of a triangulated coverage-- a triangulated irregular network or TIN-- for defining digital terrain. Each facet of this network can be defined in terms of its slope, its aspect and its elevation at its three vertexes. Whereas contours and x,y coordinates merely give locations of particular elevations, we actually develop a topological model of the surface so various spatial analysis functions can be performed (e.g., watershed models, slope calculations, sun intensity, viewshed, etc.). ESRI's TIN procedure makes it possible to neatly integrate the TIN with all of the other coverages in the GIS. The actual conversion is quite simple (with the exception of surfaces which contain major, undefined discontinuities, such as cliffs, etc.); use of a fifth order bivariate quintic surface fitting tool allows extraction of the extreme points of the surface as a set of sample points for constructing the TIN network which defines the surface. We can also use contours and random x,y,z coordinates to build a TIN and integrate it into a GIS data base.

MAP ACCURACY IMPROVEMENT TECHNIQUES

It is our experience that considerable resources are often required to standardize the information going into a GIS. Part of the reason for this is that one must be very explicit in the way the information is actually organized if the computer is to be able to accept it. To meet this high standard is often expensive because it requires cleaning up the source information (unfortunately this cost is blamed on "automation"). For example, maps going into GIS's vary in accuracy and consistency. In manual cartography the problems this causes can often be dealt with readily, but in digital mapping explicitness is required and such inconsistencies create problems, particularly when users want to overlay maps, either graphically or topologically, searching for relationships. Therefore, over the years, ESRI has developed many tools and techniques which address these problems. Some of these are described in another paper (1). These techniques fall into three categories: pre-automation techniques; during automation techniques; and post-automation techniques.

Pre-Automation

There are often conflicts between different maps in the representation of the same feature (e.g., the delineations of the aquatic vegetation of a swamp on a land cover map, and of the swamp on a habitat map). This may be because the original maps were made at different scales, different resolutions, using different classification systems; or were done at different times, with different scientific disciplines doing the mapping, etc. To resolve such problems ESRI has developed the Integrated Terrain Unit Mapping (ITUM) approach. The terrain units mapped are based on the photointerpretation of naturally occurring, visually discernable areas of homogeneous appearance (photomorphic units). (By photointerpreting recent imagery, map updating is done as ITUM goes forward). These photomorphic units are then associated with other mapped information and, if necessary, subdivided into smaller units (e.g., because soil and geology maps show that alluvial gravel deposits cross the swamp). Finally, Integrated Terrain Units (ITU's) are created such that each ITU has only a single soil type, landform, vegetation (land) cover, surficial geology, slope, surface hydrography, etc. Thus all ITU's having the same code have the same characteristics. (Such ITU's tend to be consistent "response units" for many modeling purposes.) The ITUM process retains at least the accuracy of the original maps, and usually enhances it through intercomparison of the variables mapped. As a result of ITUM a number of variables from a variety of data sources are compressed onto a single polygon overlay to the base map; this not only brings together data which, in isolation, would have less meaning, it also makes for more efficient automation and editing.   Yet the information which can be extracted from ITUM is the same as that extractable from parametric mapping; independent soils, land cover, etc. maps can be produced.

ITUM mapping has been done on all continents and at virtually all common map scales. The various scientists we work with on these projects often argue against its use until they have actually gone through the process and understood it; then, without exception, they have acknowledged its value as an integration technique.

We have also applied these techniques to integration of other cultural phenomena; we minimize the number of separate overlays which must be automated by placing compatible variables together on overlays. In Washoe County (Reno, Nevada), for example, they integrated land parcels, roads and administrative boundaries on top of a consistent basemap before they did their automation. It saved them a lot of money, and their data base is very consistent.

During Automation Techniques

We have also developed procedures that take advantage of the actual automation process to make adjustments to the data being captured. Two will be described here.

The first of these is <u>templating</u>. In templating one set of information, such as coastlines, is automated once and then used as a template when other thematic layers are automated . This saves both the time for re-automating the coastlines again and again, and insures consistent information for that particular data set. We have also used templating to put in a control grid and snapped phenomena that we are entering into the data base to that basic templated grid. We've done this, for example, to the township, section, quarter section and quarter-quarter section reference locations which have been the orientation grid for the entry of land records. Because of the topological structure of ARC/INFO-based GIS's, templating provides more than just a visual reference or just another layer within the digital data bases.

<u>On-line transformation</u> is a technique in which data are entered in coordinate measurements and transformed into ground coordinates relative to a geodetic network. The data are transformed during the digitizing effort. This simplifies the processing and creates more accurate data.

<u>Text/annotation</u> can be extracted from the tabular data base and automatically associated with map features. Examples of this automatic annotation technique are road name placement from DIME or ETAK files, automatic dimensioning of land parcels and polygon text labeling. This is considerably more efficient than the manual entry of feature annotation.

<u>Automatic snapping and "on the fly" topology construction</u> is used to immediately check closure and consistency of all map features. This provides immediate feedback to the data entry person. This functionality requires arc splitting, automatic node insertion and immediate regeneration of attributes within the data base.

Post Automation Techniques

The following are techniques that allow for information to be adjusted using various automatic or manual interactive graphic procedures:

Manual editing (interactive graphics) involves use of tools using graphic screens and cursor technology to interactively adjust cartographic features on a display using the common types of interactive tools found in a CAD technology: deleting, adding, rotating, adjusting and splining, etc..

Interactive rubbersheeting allows the user to interactively select any x,y point on a graphic screen and indicate a second x,y point to which the first is to be displaced. A whole series of such displacement points can be entered. The user then exercises a command which generates a three-dimensional adjustment surface for all points on the map. The adjustments are then made automatically, using a fifth order interpolation technique (bivariate quintic interpolation). To prevent lines from being pulled apart as rubbersheeting is performed it's essential that the data have a topological structure. We've used interactive rubbersheeting frequently in cartographically adjusting DIME files to higher quality cartographic displays illustrated on a background coverage before conflation.

It is possible to use this technique in combination with digital video images to interactively correct vector data. An integrated raster and vector terminal is used to create an image background display on a graphic screen. Pixel memory is used to display image data from LANDSAT or other photography using a background graphic plane. Simultaneously segment memory is used to display graphic vector information as a foreground graphic plane. Then, while visually interpreting the image, the vector information is adjusted to the image using the interactive graphic editing or rubbersheeting techniques described above.

Conflation is a set of procedures developed initially by the U.S. Census Bureau to transfer attributes coded in DIME files to road centerline files (which are of higher cartographic quality) such as city engineered road centerline files or DLG files being scanned by the USGS. This is an automatic tool which uses interesting techniques (similar to polygon overlay): lines are intersected and attributes associated with one line can then be associated with the coordinate strings of another line.

In the development of Integrated Terrain Unit Maps a procedure, attribute consistency analysis, was devised for checking that the attributes within the polygon attribute coding were consistent with one another. For example, it may simply not be possible to have both water and granitic outcrops as attributes of the same polygon, or, more subtly, it may be impossible to find desert soils with marshland vegetation which also have a geologic structure normally found in alpine mountain regions. We developed a knowledge-based, table look-up procedure by which all of the polygons in the data base can be checked for attribute consistency and errors flagged for correction.

Another inter-layer technique involves line snapping. Lines from one data layer can be automatically associated with lines of another layer, with all inconsistencies automatically removed. For example, land cover classes correspond directly with lines of roads; yet when these phenomena are mapped separately they can occasionally have inconsistent lines. But, using this procedure, "roads" shown on a land cover map can be "snapped" to actual road lines.

Fuzzy tolerance is a mathematical technique for snapping and generalizing points or lines which share approximately the same location, into a single point or line. We often use this technique to correct and homogenize poorly digitized or poorly captured information.

ENTRY OF TABULAR DATA

While key entry must sometimes be used for automation of attribute information, the tabular data in urban information systems are often already automated. Usually such files can easily be brought into a relational data base. More difficulty is experienced with information which has greater amounts of structure (such as a hierarchical DBMS or a network DBMS-- see Figure 5) because the tabular data must be extracted and then associated with the relational GIS information. Where cities maintain their data in hierarchical systems (IMS and the IBM system), performing such extractions and reorganizations at the time of each data base query is difficult. A sensible long term strategy is to bring all the data into a relational environment. Short term solutions are more difficult.

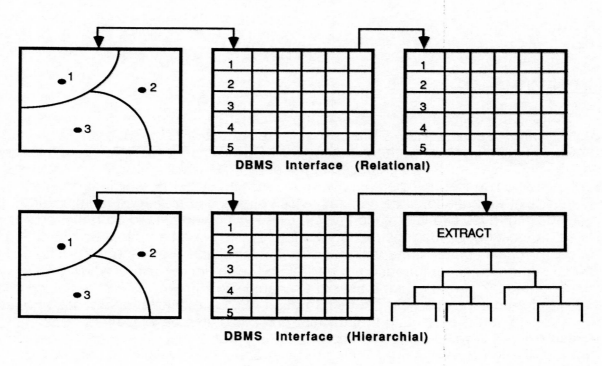

DBMS Interface (Relational)

DBMS Interface (Hierarchial)

Figure 5

ESRI has built a number of interfaces to existing tabular data bases. These have recently been generalized into a relational data base interface (RDBI) (see Figure 6). This interface underlies our GIS software and allows the user to interface multiple relational data bases; RDBI uses the SQL standard. We developed RDBI because the organizations with which we deal have different sorts of relational data bases, their investments in these are considerable and interfacing is often preferable to conversion in such cases.

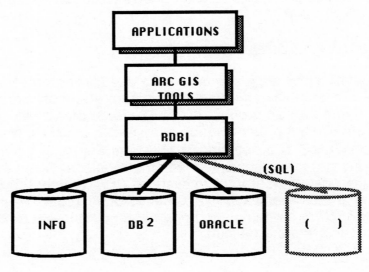

Figure 6

TEXT INFORMATION

There is a growing need to integrate text with GIS's. Some of the more modern DBMS systems, such as INFO DB and, to a lesser extent, INGRESS, are attempting to interweave text information with the data in their DBMS's. The first system of this type was the INFOTEXT system, introduced in the early 80's, which brought together two data types and two technologies (i.e., word processing systems, which stored and managed text, and relational data bases, which stored and managed tabular data). By cross-indexing these two information types, INFOTEXT was able to do rather complex queries on indexed text information. In the future, GIS's will be interfaced to relational data bases which contain large volumes of text information (See Figure 7). This will allow zoning ordinance information to be "related" with zoning codes which are, in turn, related to tax parcels; then a user query could elicit the full text of the zoning code pertaining to a particular parcel. In fact, through various relational operations and fast text-searching, maps pertaining to particular text could also be displayed. This interface exists now and will soon become part of urban GIS's. We already see some of the intelligence agencies inter-linking the fast data finder, relational data bases and maps.

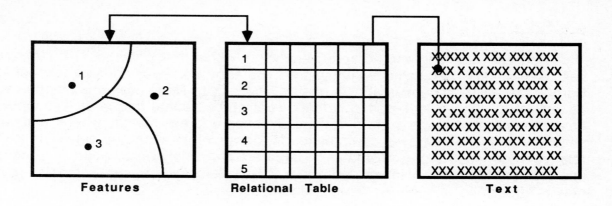

Features   **Relational Table**   **Text**

TEXT INTERFACE

Figure 7

As one way of taking graphics primitives from a GIS and integrating them with text for final output on a laser printer we are making use of PostScript.   Others seem to be adopting this same approach.

DIGITAL LASER DISK INTEGRATION

A number of organizations are experimenting with video disk and laser disk technology linked to GIS systems. The State of Wisconsin, for example, has associated its digital road maps, topologically structured and indexed by road mile, with its road photo logging system, which takes pictures in both directions at 1/10 of a mile intervals for the entire state road network. This integration allows analysts to point at a road location on a map and almost instantaneously look in both directions at the visual characteristics of that particular road (see Figure 8). By integrating address information into the same data base, one can simply type in an approximate road address or street address and see the pictures which are appropriate to that particular location. This kind of integration will have multiple applications, including accident investigation, road condition evaluation, and sign evaluation; even taking visual trips down routes automatically selected to meet given modeling criteria.

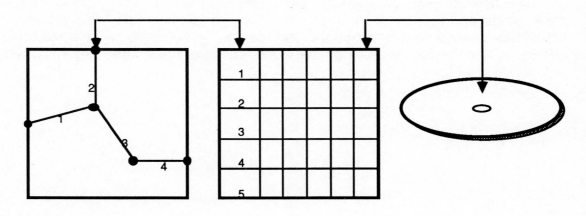

GIS Interface to Pictures

Figure 8

ADDRESS MATCHING

Address matching and address geocoding, within the context of GIS technology, has been going on for some twenty years. The early work was done by the Census Bureau for the 1970 census and its related DIME files. Now, newer techniques such as knowledge-based look-ups, automatic spelling correction and related technologies, have made it possible to perform rather advanced address matching with minimal difficulty. Modern toolboxes can now readily parse information and quickly index it relative to geocoded road segments. These are not simply batch address geocoding operations; they are fully integrated with a GIS system, thereby developing all sorts of new relationships among the various address locations.

CONCLUSIONS

The following are some of the conclusions which might be drawn from the above.

GIS is becoming a tool which can interface with a lot of other technologies (see Figure 9): computer-aided drafting technology; remotely sensed information; tabular data bases containing descriptions of geographic features; photo logging technology; video and laser disk technology; word processing systems; address information systems; terrain modeling systems; legal record systems; manual maps in various sizes, shapes, forms, scales, transformations, projections, etc.

Making the interface to all of these various data types is not a simple conversion operation. A variety of software and procedural tools are necessary to craft the information into a consistent and integrated geographic information system format. Nevertheless, geographic information in these other formats can be productively used when integrated into a common GIS data base. The synergistic value from this cross-indexing of information in the GIS is substantial.

While considerable discussion has taken place about the need for greater standardization, tools and technologies which bypass the need for standards have been developed. ESRI's ARC/INFO, for example, has some 15 different data base interfacing tools and procedures; these take information in different structures, reorganize it and integrate it into a single, standardized data base.

The common geo-index which interrelates the various text, graphic, cartographic, image, photo and tabular information can be extremely valuable. GIS technology lets us navigate through and between the various information media. This will have profound effects on how we design the workstations of the future: they will have to be

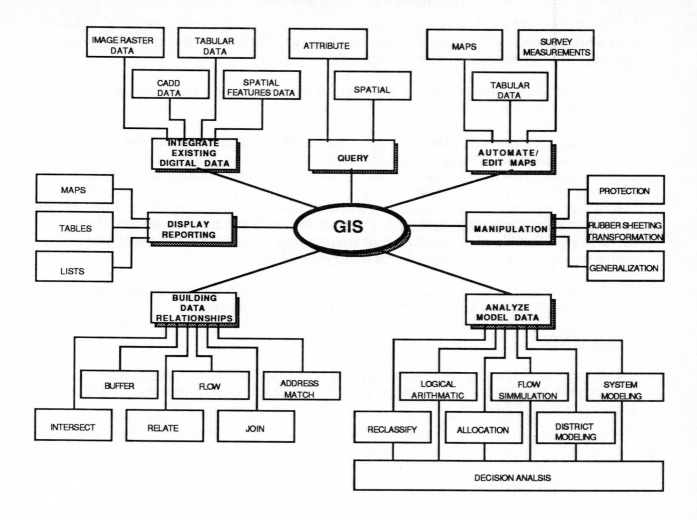

Figure 9

responsive to all these media and data types; have to handle digital images, photos, text and tabular data; and will have to have operators and query tools that allow the mixing and matching of all these. This may mean either multiple screens or just large screens which integrate images and vector information, like those on the more modern workstations .

To allow the typical professional to easily access and use these capabilities, a better human/computer interface will also be needed; simpler tools, like mouse-selected, pull-down menus, will have to be implemented. No doubt this will eventually involve two-way voice communication as well.

Network architectures with high-volume, high speed communication capabilities and workstations which can integrate all of these different data and media types will become the hardware standard in this field.

Given all these capabilities, we need to work harder on how to effectively share information among our institutions.

Finally, it's obvious that these developments are changing the way some of our institutions function, changing the institutions themselves, and probably changing the way we think. Are these changes user driven-- or just technology driven? Can we adapt all this technology to real needs, or are we merely adapting to the technology? We're having a hard time introducing even simple GIS technology into cities; maybe we're just surpassing the capabilities of people and institutions to absorb this new technology. But it's also possible that we're trying to introduce the technology into some situations where it doesn't-- at least yet-- belong.

Given that possibility, we need to be careful that we are not overwhelmed by approaches that have little to offer except the latest technology.

REFERENCES

(1) Dangermond, Jack, Bill Derrenbacher and Eric Harnden. "Description of Techniques for Automation of Regional Natural Resource Inventories." 1982. (Available from ESRI.)

SECTION 3
Data Manipulation

Overview

This section consists of three articles concerned with fundamental GIS operations. Robinove describes methods of spatial analysis in the context of the rules of logic and spatial feature relationships. Berry, in the second paper, discusses fundamental methods of cartographic modeling and spatial statistics. The third article, by Stefanovic and Sijmons, describes techniques for the representation of topographic relief through digital elevation models.

Suggested Additional Reading

Hopkins, L.D. 1977. Methods for Generating Land Suitability Maps: A Comparative Evaluation. *American Institute of Planners Journal*, 43(4):386–400.

Lupien, A.E., W.H. Moreland, and J. Dangermond 1987. Network Analysis in Geographic Information Systems. *Photogrammetric Engineering and Remote Sensing*. 53(10):1417–1422.

Lyle J. and F. P. Stutz 1983. Computerized Land Use Suitability Mapping. *The Cartographic Journal*, 20(1):39–49. Reprinted in William J. Ripple (ed.), *Geographic Information Systems for Resource Management: A Compendium*. 1987. Falls Church, VA: American Society for Photogrammetry and Remote Sensing and American Society on Surveying and Mapping, pp. 66–76.

Principles of Logic and the
Use of Digital Geographic Information Systems

By *Charles J. Robinove*

ABSTRACT

Digital geographic information systems allow many different types of data to be spatially and statistically analyzed. Logical operations can be performed on individual or multiple data planes by algorithms that can be implemented in computer systems. Users and creators of the systems should fully understand these operations. This paper describes the relationships of layers and features in geographic data bases and the principles of logic that can be applied by geographic information systems and suggests that a thorough knowledge of the data that are entered into a geographic data base and of the logical operations will produce results that are most satisfactory to the user. Methods of spatial analysis are reduced to their primitive logical operations and explained to further such understanding.

INTRODUCTION

The purpose of this report is to explore, in a philosophical sense, the creation and use of digital geographic information and systems from the standpoints of the creator and the user. It is vital to the proper operation of geographic information systems that basic principles of logic be followed, that the limits of what a system can or cannot be expected to do be well understood, and that the relations, both logical and spatial, among the various data elements be understood.

To avoid confusion, two terms need to be defined. These terms have been used in a somewhat loose manner in the literature. For a precise understanding of this report (although some may disagree with the specific definitions), the terms are defined below.

Geographic data base. A collection of digital map data in which each feature is referenced to a geographic location expressed in spatial coordinates. Data may be points, lines, or areas. The data base is analogous to a single map or a set of map overlays. It may also be referred to as a "spatial data base."

Geographic information system (GIS). A collection of computer programs in a given hardware environment which operate on a geographic data base to analyze individual data-base elements or for synthesis of multiple data-base elements. A GIS takes into account the spatial position of each element as well as its other characteristics.

This definition includes image-processing programs and computer-aided mapping programs as well as software packages that are specifically designated by their developers as "geographic information systems." A digital geographic data base can be visualized by an analogy with a series of map overlays. McHarg (1969) has successfully used a map-overlay system to display the common attributes of selected areas in order to make decisions on the type and degree of land development that is commensurate with the physical properties and limits of an area. A digital geographic information system may use the same data as a map-overlay system, but because the data are in digital form, the system is much more flexible in the type and amount of the data it can handle, in the logical operations that can be performed on the data, in the ease of changing or updating the data, in the forms in which output products can be produced, and in the statistical generalizations of the data that can be made.

This report does not discuss the algorithms or computer programs and operations that are required for creation and analysis of a geographic data base. It does, however, attempt to place those operations within a rigorous philosophical framework so that the reader can appreciate what he is asking of the system and what he can expect to get out of it.

Published by U. S. Geological Survey, 1977.

The report by Calkins and Tomlinson (1977) is an excellent general guide to the creation and use of geographic information systems. It does not describe any particular system, but it describes practical methods and constraints for system design and use.

DESIGNING A GEOGRAPHIC DATA BASE

Data that are selected for use by a GIS may be "raw" data in basic form or data that have been processed, mapped, or interpreted in various ways. The creator of the data base must make a number of choices involving tradeoffs among completeness and detail of the data sets to be analyzed, the amount of data manipulation that must be done to answer a user's question, and the amount of detail that is required by the user.

Suppose that a data base is to be designed for evaluating streamflow in the United States. It would be possible to enter mean, minimum, and maximum daily flows for more than 10,000 stream-gaging stations in the United States, some of which have more than 50 years of record. Such a data base would be very large and would be expensive to use. A simpler data base would contain maps showing statistical generalizations of the streamflow characteristics by drainage basin, such as mean, minimum annual, and maximum flow. This data base would be simple and inexpensive to use to answer general questions, but its information would be usable only on a national or regional comparative basis and would not be capable of responding to a user's detailed question about streamflow at a particular point.

We cannot expect any data base to be capable of answering all questions (an old proverb states, "A fool can ask questions that wise men cannot answer"). We can, however, expect that intelligent decisions (and even guesses) can be made as to what questions a data base would be asked. Intelligent decisionmaking requires a thorough knowledge of the actual or potential user community and its interests.

It is customary to visualize a GIS as analyzing a series of data planes, with the capability of analyzing data in a single plane and also of showing the relations among selected sets of those data planes. Data in a single plane may be raw data or the result of previous processing, or they could have been created by the GIS. For example, a data base might contain a plane consisting of digital elevation data. From this layer, additional maps can be constructed, for example, maps showing slope, rate of change of slope, and aspect.

LOGICAL AXIOMS FOR DATA USED BY GEOGRAPHIC INFORMATION SYSTEMS

Geographic data bases contain individuals, classes, attributes, and statistical and mathematical generalizations of attributes. An individual is a single data point, such as a value at an x, y coordinate, that cannot be logically divided. A class is a collection of individuals. The individuals may have the same value and be at different locations, or they may be at the same location and have different values. An attribute is any value, quality, or characteristic that belongs to an individual or a class. It may be a name, a numerical value, or a statistical parameter.

Data of any type can be statistically characterized, but in a geographic information system attributes characterize a location or an area because spatial position is of equal importance to the value of the attribute, and indeed is a unique characteristic of the individual.

Analysis of data in a geographic information system requires that the user recognize some basic logical axioms that apply to the data and their relations. Most of the axioms can be considered common sense, but it is worthwhile to make them explicit for complete understanding of the data, their attributes, and their relationships.

A data plane is a collection of features with x, y coordinates; the attribute forms the z-coordinate. The data plane may be coded and displayed in either vector or grid-cell form. In vector form, the data are manipulated and displayed as points, lines, and areas. In grid-cell form, the data may be visualized as cells of a regular grid, each cell having a value. A data plane may show values in nominal, ordinal, cardinal, or ratio form.

A point is the smallest mappable unit to which a property may be attributed. A line is the locus of all points that have the same attribute and within which no point is adjacent to more than two other points. An area is the locus of all points having the same attribute and within which any point may be adjacent to three or more other points.

The four axioms on which our argument rests are

1. There exists at least one individual (that which exists as a separate and distinct entity). If this were not so, there would be no subject to discuss.

2. The individual possesses at least one attribute. If this were not so, it would not be an individual.

3. Individuals are distinguishable from each other on the basis of their attributes. For example, there may be many houses in an area, but they are distinguishable on the basis of attributes of size, value, number of occupants, and so forth.

4. Individuals and attributes may be classified into usable categories. The normal principles of logic and symbolic logic are based on the above axioms but do not usually take into account the statistical, temporal, or spatial relations of individuals. The field of geography has implicitly used the principles of logic with relation to place but has not explicitly formulated the rules and principles that allow logical operations to be performed on spatially distributed data. Nystuen (1968) has identified three fundamental spatial concepts in the development of an "abstract geography" but urges that empirical work in the field of geographic analysis remain strong. The three concepts are direction, distance, and connectiveness. Nystuen points out (p. 39) that connectiveness is a topological property of space and that it is independent of direction and distance. All three properties are needed to establish a complete geographical point of view. The concept of connectiveness subsumes the concepts of adjacency, proximity, superposition (vertical connectiveness), and containment.

A number of further axioms are given below with an example of the application of each. The term "data" refers to an attribute of an individual as represented in the spatial data base. The term "x, y coordinate" refers to the spatial position of that individual in the data base. The "individual" may be a point, a line, a grid cell, or a polygon. It is the smallest homogenous unit to which attributes may be assigned.

1. Data at the same x, y coordinate in all data planes apply to that x, y coordinate (universal).

2. Data at an x, y coordinate in some data planes may be valid for that x, y coordinate and for some region around that coordinate; the region of influence varies as a function of a radius, a numerically defined region (such as a Thiesen polygon), or a spatial frequency of occurrence.

3. Data at an x, y coordinate in some data planes may be valid for that x, y coordinate and for some region defined by a boundary in another data plane (attribute within a region).

4. Data along a straight or curved line of x, y coordinates may be valid for those coordinates and for some region whose boundaries are parallel to that line; the region of influence varies as a function of the distance (proximity to a road or to the centroid or edge of an urban area).

5. Data at an x, y coordinate in some data planes may be valid for that x, y coordinate, and their relation to the immediately surrounding x, y coordinates is defined in only one or two directions (maximum slope of a surface).

6. Identical data at various x, y coordinates in a single data plane are identical (universal).

7. A feature in one data plane may be used to select data within its x, y coordinates from other data planes (masking or "cookie cutter").

8. Varying data may be replaced by uniform data within a boundary (statistical classification or generalization).

9. Data at an x, y coordinate may be characterized by their similarity to or difference from data at neighboring x, y coordinates (filtering).

10. Data at an x, y coordinate or aggregations of data at various x, y coordinates may be named (labeling) or ranked.

11. Statistical measures and procedures may be applied to aggregations of data regardless of their x, y coordinates (classification) or to data in different layers at the same location (correlated layers).

12. Data in a neighborhood may be characterized by their spatial relation to a point within or exterior to the data plane (aspect of a surface or intervisibility).

13. Data having uniform attributes may be counted (number, length, area, or spatial frequency).

WHAT IS A CLASS?

The logic of classes is treated in Werkmeister (1949) and Carnap (1958) as an extension of the calculus of propositional functions. However, in formulating and using the traditional rules of symbolic logic, classes, their values, and the relationships among classes consider only their attributes or functional relationships and not their spatial relationships. The traditional rules of propositional calculus and symbolic logic apply to the creation and operation of geographic information systems, but, in addition, the attributes of classes may depend on their spatial shape, their connectiveness to other classes, or to a statistical measure of the attribute of a single class or a group in close spatial proximity.

The principles of symbolic logic in a geographic information system are most efficiently applied to classes of points, lines, and areas. A class is usually thought of as a group of points that may be spatially coextensive and that have the same attributes. However, it is possible, and in many cases quite necessary and useful, to define as a separate class all of the spatially coextensive points in a given class that are within a certain spatial distance from the spatially coextensive points in another class. This can be illustrated by considering the edge of a landmass at the ocean. The land is one class and the ocean is another class. But it is possible to postulate a third class—the shoreline, which is a unique class of coextensive points. We may then define an additional class of land that is within a specified distance of the shoreline, thus creating four useful classes when we started with two. This process can go on indefinitely. Mandelbrot (1977) showed that the class of shorelines is a fractal (a curve without a tangent); that is, between any two points along the shoreline, the line itself may be as short as the shortest distance between the two points or it may be of almost infinite length, depending on the scale and resolution at which it is mapped and portrayed. The question of scale of portrayal of the classes considered by Mandelbrot is relevant to geographic information systems when one is deciding at what map scale to analyze and display the information. It will suffice to say that the length of the line defining the boundary between two classes (or the line defining a class itself) is dependent on the scale of mapping (in cartographically drawn maps and in digital data displayed in a vector format) or on the resolution of the data (for displays of grid-cell or raster-format data). Thus, the definition of a class may depend not only on the attributes of its members, but also on their distribution: a class may be defined as (1) a point with one member and one or more attributes, (2) a line with a number of spatially aligned members, no one of which is adjacent to more than two other members and which must have one or more attributes, and (3) an area with

three or more contiguous members which must have one or more attributes.

A geographic data base contains classes with what are usually referred to as "x, y, z coordinates." X and y are Cartesian coordinates which may be related to map coordinates in a given projection, and z is the value of an attribute. All data planes have the same system of x and y coordinates, and each one shows a class or classes, each with one or more attributes.[1] The classes may be indicated on the data plane or described in separate files.

In the strictest sense, any x, y coordinate has only one z-coordinate or attribute in a single data plane. Common sense tells us, however, that it may have many attributes. For example, the attribute "forest" has a high correlation with "on land," "presence of leaves," and "ability to transpire water."

It is the problem of covariance of attributes that occupies the interpretive mapper and is of particular importance when there is a spatial covariance to be described as well as an attribute covariance at a point.

THE RELATIONS AMONG CLASSES

The relationship of classes is based on the calculus of propositional functions, with the addition of the spatial-class definitions given previously. The following is paraphrased from Werkmeister (1949), as an explanation of the postulates governing class relations.

Definitions
1. Universal class=everything=1.
2. Null class=nothing=0.

Postulates
1. If a and b are classes, then there is class a+b (logical sum). The logical sum consists of the members that are distributed between a and b. This is the logical "or" operation.
2. If a and b are classes, then there is class a×b (logical product). The logical product of two

---

[1]There are two other specialized systems that should be mentioned but that are little used at present. The first uses stereoscopic parallax as a fourth coordinate, which allows display in a three-dimensional view of the other three coordinates—two spatial and one an attribute. This has been used, for example, to display magnetic intensity superimposed on a color display of a satellite image. The second involves animation of change in an attribute or in the position of an attribute and is best displayed as a motion picture or video display. A third system could be postulated which would combine both of these, thus giving a three-dimensional animated color motion picture of attribute change. I know of no system that does this at present.

classes consists of the members that are common to a and b. This is the logical "and" operation.

3. There is class 0 such that a+0=a, for any class a. The logical sum of class a and the null class is identical with the class a.

4. There is a class 1 such that a×1=a, for any class a. The logical product of class a and the universal class is identical with the class a.

5. If there is a universal class 1 and a null class 0, then for any class a there is a class ~a, such that a+~a=1. The logical sum of a class and its negative exhausts the universe of discourse, whereas a class and its negative exclude each other, that is, a×~a=0. This is the logical "not" operation.

6. If a, b, a+b, and b+a are classes, then a+b=b+a. This is the Commutative Law for Logical Sums.

7. If a, b, a×b, and b×a are classes, then a×b=b×a. This is the Commutative Law for Logical Products.

8. a+(b×c)=(a+b)×(a+c). This is the Distributive Law for Class-Sums.

9. a×(b+c)=(a×b)+(a×c). This is the Distributive Law for Class-Products.

10. There are least two classes, a and b, such that a≠b, that is, a is not identical with b.

The relationships defined in these postulates can be clarified by the diagram in figure 1.

- Let the rectangle represent the universal class 1.
- Everything inside the complete circle "a" is included in the class a.
- Everything inside the complete circle "b" is included in the class b.
- Everything outside the circle "a" but inside the rectangle is ~a.

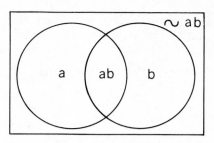

FIGURE 1.—Schematic diagram of two classes in map form.

- Everything outside the circle "b" but inside the rectangle is ~b.
- Everything belonging to the total area of the two circles is included in the class-sum a+b.
- Everything in the overlapping area of the two circles is included in the class-product a×b.

By shading the area of the null class, we can illustrate the product of a class and the null class, as shown in figure 2.

Other relations and functions can be represented in a similar manner. Once the postulates have been given, it is possible to introduce the relation of class-inclusion by definition. Accepting "⊂" as the symbol for class-inclusion, so that "a⊂b" means "class a is included in class b," we define

1. (a⊂b)=(a×~b=0)
2. (a⊂b)=(b+~a=1)
3. (a⊂b)=(a×b=a)
4. (a⊂b)=(a+b=b)

These definitions, in conjunction with the postulates, enable us to derive an indefinite number of principles or theorems of the class "calculus." The following examples are self-explanatory. (We simplify by writing ab instead of a×b.) The symbol "⊃" means "implies."

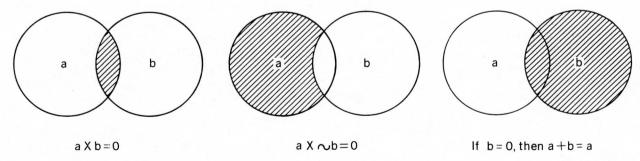

a X b=0   a X ~b=0   If b = 0, then a + b = a

FIGURE 2.—Venn diagrams of the logical products of two classes. (From Werkmeister, 1949, p. 435.)

1. $ab \subset (a+b)$ (from postulates 2 and 1)
2. $ab \subset a$ and $ab \subset b$ (principle of simplification)
3. $[(a \subset c) \cdot (b \subset c)] \supset [(ab \subset c)]$
4. $[(a \subset c) \cdot (b \subset c)] \supset [(a+b) \subset c]$
5. $[(a \subset b) \cdot (a \subset c)] \supset (a \subset bc)$
6. $[(a \subset b) \cdot (a \subset c)] \supset [a \subset (b+c)]$
7. $[(a \subset b) \cdot (b \subset a)] \supset (a=b)$
8. $a0=0$
9. $[(a \sim b=0) \cdot (b \sim c=0)] \supset [(a \sim c)=0]$
10. $[(ac=0) \cdot (bc \neq 0)] \supset [(b \sim a) \neq 0]$

Although the Venn diagrams used by Werkmeister (fig. 2) are meant to illustrate symbolically the logical relations among classes, they can easily be used to illustrate also the spatial relations among classes. If we consider figure 3, we can visualize the rectangle as the base map (or base data plane) in a geographic information system. In that figure, a is an area class while the map area outside the class is null. It is then easy to see that the methods of overlaying mapped classes in separate data planes is an exercise in logic which is carried out by Boolean logic operations in a computer following the previous axioms and the postulates of symbolic logic. These axioms and postulates are usually described as the logic of functions "and," "or," and "not." The function "and" applied to a and b in figure 2 would result in a map of the total areas of a and b; the function "or" would result in a map of ab; and the function "not" would result in a map of the areas of a and b that are not common to each other. These areally coextensive classes may be represented in a geographic data base either by vectors showing the boundaries of the classes or by grid cells showing the area of the classes. Conventionally, "+" means "or," "×" means "and," and "~" = not.

The "and" function operates in a computer system according to the following truth table:

If A and $B=C(A \times B)$,

A	B	C
0	0	0
0	1	0
1	0	0
1	1	1

The "or" function operates according to the following truth table:

If A or $B=C(A+B)$,

A	B	C
0	0	0
0	1	1
1	0	1
1	1	1

The "not" function operates according to the following truth table:

If not $A=B$ $(\sim A=B)$,

A	B
0	1
1	0

These functions can be concatenated into more complicated but useful ones, such as the EQV functions:

$$C=(A \times B)+\sim(A+B)$$

A	B	C
1	1	1
1	0	0
0	1	0
0	0	1

A Venn diagram can be applied to spatial data without consideration of the spatial relations to promote understanding of the possible and impossible (or unlikely) combinations of data. Varnes (1974) demonstrates in his figure 23 the relation among slope, firmness, and thickness of geologic units in order to determine their suitability for engineering purposes. This figure, and indeed his entire report, should be consulted for a rather full explanation of map logic and the attributes of map units. His explanation of the logic of map units is based largely on areally coextensive units and only slightly on point and line data, which are of major importance in digital geographic systems.

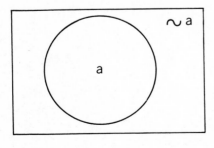

$$a \times 1 = a$$

FIGURE 3.—The null class.

The logical relations of spatial position (fig. 4) may be explained as follows:

1. Adjacency: Class A is always adjacent to class D (for example, water is always adjacent to land).
2. Proximity or connectedness: Class A is always within a certain distance of class D but is never adjacent to it (for example, end members of a continuum, such as vegetation density).
3. Superposition (z direction): Class A always lies above class D (for example, one layer of rock always overlies another layer).
4. Containment: Class A always lies within class D (for example, the hole in the doughnut is always surrounded by the doughnut).

A classification must be logical. For example, one cannot logically establish classes for a mapped area such as (1) wooded, (2) urban, (3) recreational, because a wooded site may or may not be recreational and an urban site may or may not be wooded. It would be necessary to classify each area of the map as wooded or not wooded, urban or not urban, and recreational or not recreational and then use logical rules to show the relations of the three maps.

Varnes (1974) notes four operations that are performed on maps: generalization, selection, addition and superposition, and transformation. Each of these operations must conform to the log-

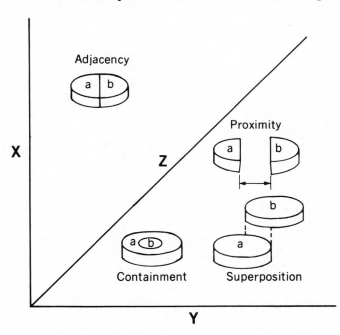

FIGURE 4.—Logical relations of spatially located classes.

ical principles previously outlined. The performance of these operations on digital data is done by computer programs. An important point to keep in mind is that some of these operations can be performed on data before they are put in map form, such as tabular attribute data sorted and categorized by a data-base management system. A decision must be made for a given area as to what to map and what not to map. This is a process of selection. A decision must also be made, for any mappable unit, as to what attributes to describe or measure in detail and which to infer on the basis of attribute covariance and which not to. This is the process of generalization. It is not possible for the individual who makes an original map (from detailed field data) to predict all the questions the user may eventually ask about the area. He can only map, to the best of his ability, the features he believes will be useful; in the mapping process, he selects and generalizes. The product of his field mapping becomes the input to a digital data base. If his map has detailed quantitative data, the digital data base will be detailed and quantitative. If not, it won't. If a severe process of generalization has occurred before the map is digitized, the original data cannot be recovered and the map may be useful for only a single purpose. An extreme example would be a set of map data on soils, soil permeability, slope, depth to the water table, and other factors used to create a "stoplight" map of suitability for waste disposal (that is, the map would indicate that a location either is or is not suitable.) If the stoplight map is the only map stored in the data base, it is usable only for the purpose for which it was designed; it cannot be decomposed into its parametric attributes. If, however, the basic attributes are stored at the level at which data are collected in the field, the final interpretive map can be recreated with the same or different generalization criteria and, in addition, other derivative maps can be created from the same basic data for other purposes. For this reason, it is always desirable to store the data in as primitive a form as possible. Derivative maps can also be stored for later use, but they should supplement, not replace, basic data. The processes of addition and superposition follow the logic of the Venn diagram previously explained. One map can be superimposed on another, a program can select the areas of common attributes from two maps, maps can be added together, and a program can select only the minimum or maximum values occurring in two maps.

Maps may be transformed by changing the character of their symbols or the means of presentation. The original data, such as a set of points with values on a digital map (for example, elevations on a geologic horizon), may be contoured or a mathematical surface may be fitted to map the elevation of the horizon at any point in the map. Some transformation operations may be performed consistently and efficiently on a digital map to create entirely new maps. A map of roads may be operated on by a proximity mapping program to display all areas that are within a given distance of the roads. This proximity map may then be used as a mask to show only certain features in the proximity area. The use of a Venn diagram can be illustrated by an example from the field of ground-water analysis. In a coastal area, high pumpage of freshwater from an aquifer will result in a decline of water levels in the aquifer and may result in drawing saltwater into the aquifer. The situation may be illustrated by a Venn diagram (fig. 5). Three conditions—well pumpage, water-level decline, and salt concentration—are each described in two classes, low and high. The relations of the two conditions of the three classes are shown by shading. The classes in this case obviously have a high covariance, but the degree of covariance will ultimately depend on their spatial relationship.

The two extreme cases are (1) when pumpage is low, water-level decline is low and the salt concentration of the water is low (striped area), and (2) when pumpage is high, water-level declines are high and the salt concentration is high (black area). The latter situation occurs if wells are near the coast, where it is relatively easy for seawater to gain access to the aquifer. Section A (pump ×decline× ~salt) represents an area where high water pumpage causes high water-level declines but that is far from the coast so that there is no increase in salt concentration. Sections B (pump×salt× ~decline) and C (pump× salt×decline) are null classes; that is, they cannot exist because pumpage cannot result in a high salt concentration without high water-level decline (B) and there cannot be a high water-level decline and high salt concentration without high pumpage (C). Pumpage may be represented on a map at one or a number of points, and water-level declines and salt concentration may be represented by areas whose size, shape, position, and boundaries depend on the influence of the point pumpage sources. This illustration, then, shows the importance of both the logical and the spatial relationships of data and classes that can be used in geographic information systems.

Each of the three attributes in the above example is classified into only two categories, high and low. Either the person who puts the data into the system or the user of the data must decide on the number of categories and the boundaries between them. This is where the trouble really lies—in defining categories in a continuum of data points. The water-well pumpage in the example may range from 10 to 5,000 gallons per minute. Is the category "high" to be above 100, 1,000, or 3,000 or some other number? The same principles and questions apply to the water-level decline and the salt concentration. Thus the boundaries of each of the data sets (and their intersections) are really fuzzy and of various widths, not discrete lines as shown in the diagram. Geographic information systems operate best on discrete data in well-defined categories. It is the task of the analyst to determine how the original data can be best categorized when it is entered into a data base. In many cases, categories are either universal classes or null classes. For example, one category of land use might be "forest land." In one data plane of a geographic data base, an area would be displayed as either forest or nonforest. For many purposes it might be more useful to have the data classified by percentage of forest in grid cells or polygons by 10-percent intervals. This would allow some manipulation around the mean of the forest cover, and displays could be made of the standard deviation from the mean or other useful measurements. Jupp and Mayo (1982) present an example from Landsat image analysis in which

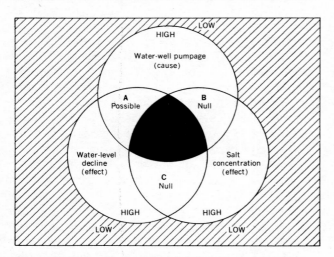

FIGURE 5.—Venn diagram showing relations among water-well pumpage, decline of water level, and salt concentration of water.

an image is classified into several categories and each grid cell is then classified further into one of several categories representing its deviation from the mean of its class. This highly useful display allows the heterogeneity of individual classes, as well as their mean attributes, to be portrayed.

These problems of classification, categorization, and boundary definition may be solved (or at least better understood) by considering the classes we deal with as fuzzy sets rather than as examples of two-valued logic or Bayesian probability statistics. Fuzzy set theory is, at the time of this writing, the subject of more than 2,000 papers and books since Lofti Zadeh introduced the subject in 1965 (Zadeh, 1965). I do not intend here to exhaustively discuss the theories and applications, but simply to indicate the basis of fuzzy logic and to suggest that it may become a powerful tool both in the analysis of spatial data and in our understanding of the data and their relationships.

The term "fuzzy system" may be considered to cover the "whole field of imprecisely described systems" (Negoita, 1981). A key to the understanding of fuzzy systems is the word "described." It implies that man's understanding of a system is as important as the precise measurement of individual observations and parts of a system. Imprecision may be a function of measurement accuracy, or it may be a function of the description of the system. A particular feature or concept may be represented by a fuzzy set. Paraphrasing Negoita (1981), we may select the concept of depth of a water body to illustrate the thinking. Let [0,1] act as a unit interval which is a set of real numbers between and including 0 and 1. We may select water at the surface to be represented by 0 and a depth of 1,000 feet to be represented by 1. We may now try to assign a specific depth measurement, say 195 feet, to a portion of the [0,1] interval with the fuzzy terms "very shallow," "moderately shallow," "shallow," "moderately deep," "deep," and "very deep." This is done by assigning a given measurement a *membership* in the [0,1] set. This is quite distinct from the *probability* relationship. This way of thinking and analysis allows us to cope with such fuzzy concepts as near and far, tall and short, steep and gentle (slopes), high and low, and old and young. The fuzzy system concept then allows us to cope with logical relations (which themselves may be fuzzy sets) in a manner quite different than the manner required with Boolean logic, which demands a two-valued logic in which a given observation either is or is not a member of a given set.

The fuzzy system concept may be illustrated by returning to the previous example of ground-water analysis in which water pumpages, water-level decline, and salt concentration were defined as either high or low, with fixed (but numerically undefined) boundaries. It would now be possible to define the grade of membership of any given observation of each of the three factors in the fuzzy set from high to low. This concept of the fuzzy set and the membership of an object (or a measurement or relation) allows both perception and linguistic description. We can then operate with a description of moderately high pumpage or fairly low water-level decline without the need to specifically and numerically define the values of each. The importance of this concept and method of operation can be seen when we must deal, in a geographic information system, with questions that are not precise (although the data may be highly precise). With the proper data base, we could ask such questions as "Where are all the areas where a geologic formation subject to landslides occurs near roads?" to determine where there is a risk of landslides damaging highways. Figure 6 is a simple example of this type of analysis. By producing a corridor along roads, dividing that corridor into arbitrary but fuzzy levels of proximity to the roads, and applying the gradient corridor only to the landslide-susceptible formation, the result is a map that qualitatively expresses the hazard to the roads.

The use of fuzzy sets and fuzzy algorithms should be considered for application to geographic data bases and should also be considered for inclusion in artificial intelligence methods for geographic information systems. Much additional research is needed on this subject.

Two other common situations occur in map analysis by a geographic information system that are not as easily handled as is the case of areas with attributes. They are the cases of point data and line data. A common example of point data is a water well for which a number of attributes are available such as its depth to water, the elevation of the water surface, the depth to the bottom of the waterbearing formation, the saturated thickness of rock, and numerous measurements of dissolved constituents. How can these attributes be mapped? Figure 7A illustrates one common method, that of constructing a Thiessen polygon on the basis of an irregularly spaced set of data points (wells) shown on a map. Lines are drawn from a well to each of its neighboring wells, the lines are bisected, and the midpoints of the lines

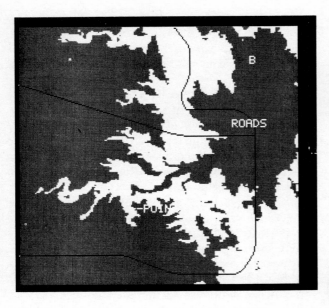

A. A hypothetical map shows two geologic formations (A and B), roads, and a point. Formation A is susceptible to landslides; formation B is not. The landslide danger to the road is inversely proportional to the distance from the road. This distance in formation A is to be mapped.

B. A corridor 16 pixels wide is mapped around the roads and the point. This is a normal mapping method, but it assumes that features within the corridor are constant.

FIGURE 6.—Gradient proximity mapping and masking.

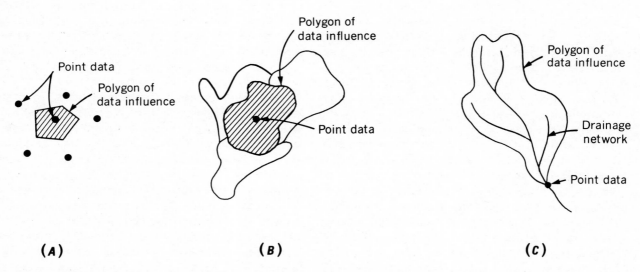

(A) **(B)** **(C)**

FIGURE 7.—Spatial representations of the influence of point data. *A*. Data at irregularly spaced points may be converted to polygons by the Thiessen polygon method. Each point then represents the value of the polygon. *B*. A measurement within a polygon whose perimeter is described by other measures may represent the entire polygon area. A point measurement may also represent the mean (or other statistical measure) of data of a mass of points in the polygon. *C*. A point measurement at the edge of a polygon represents the entire polygon, for example, a streamflow measurement representing the drainage basin. This is a special case of (*B*), and the data may be realigned to a point at the centroid of the polygon for subsequent analysis.

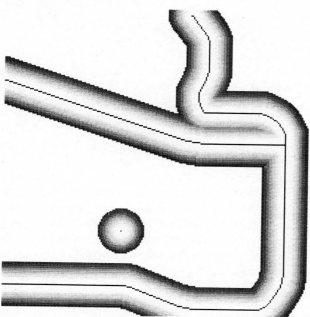

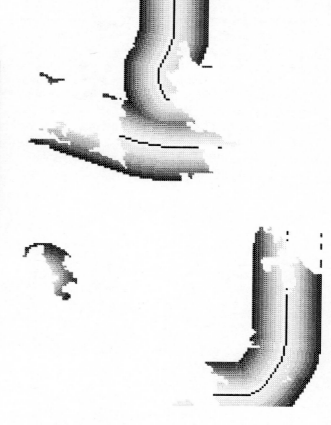

C. A corridor is mapped along the roads and around the point to a distance of 16 pixels. The brightest area, adjacent to the roads and point, is the area of maximum hazard. Away from the roads and point, the corridor darkens to the edge of the hazard zone, the area of least hazard.

FIGURE 6.—Continued

D. The hazard zone is used to mark only the map area of formation A and shows the gradient of the danger zone only where formation A occurs.

are connected to form a polygon, which is considered to be the area of influence of each individual well. The polygon can then be considered an areally coextensive class with uniform attributes. The area can be represented in a geographic information system either by vectors showing its boundaries or by a grid-cell array showing its area. An alternative method is to select a single attribute of interest, such as the elevation of the water surface in the wells, fit a selected surface to the network of data points, and represent the surface as a set of grid cells, an array of points, or polygons.

Point data within a polygon surrounded by previously described boundaries in another data plane may be easily assigned to the entire polygon. An example would be a county boundary with the amount of water consumption measured in the county assigned as a value to the entire county, as shown in figure 7B. A special case of the point-data polygon assignment occurs when the point data are on the boundary of the polygon, as shown in figure 7C. This occurs where data are taken at a stream-gaging station on a drainage network and the data at the point represent the integration of upstream attributes throughout the drainage basin. Thus, the mean annual flow in cubic feet per second per square mile can be assigned to the polygon of the upstream basin (or to its centroid) and the basin can be distinguished from adjacent drainage basins. Assigning data in this manner can lead to illogical boundary conditions, such as a sharp break in runoff per unit area, which does not really occur in nature. On the other hand, assigning the data to the centroid and then contouring can cause loss of mass balance at the gaging station.

The most difficult type of data to handle in a geographic information system are the data represented by a line with changing attributes along its length. A common situation is shown in figure 8. Data were collected at a point along a line, in this case, along a reach of a main stream between two tributary inflows. A system must be capable of assigning the data at the point to the entire line but not to the bordering areas. This can be readily handled when the line is represented by a series of labeled vectors, which really means labeling a set of shorter lines with discrete but

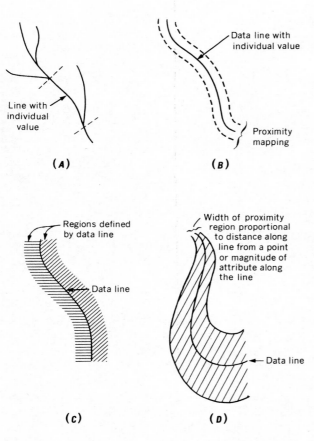

FIGURE 8.—Spatial representations of the influence of data represented by a line. *A*. A point may represent data influence between two points along a line, such as a stream, but may not be valid for any bounding region. *B*. The line may be mapped an equal distance on each side to show that its influence extends to that width. *C*. A line may define a region on either side, such as "more than" or "less than" the value of the line. *D*. If the line has a varying magnitude in a direction, the proximity (or influence) may be proportional to the magnitude of the influence.

differing attributes. In the case of a grid-cell system, the network line is composed of a contiguous line of grid cells and the aggregate of the grid cells is labeled as a class with the attribute of the data point. A second situation is illustrated in figure 8*B*; in this situation, proximity mapping creates a class bounded a given distance from a line, such as mapping of all the areas within 1 mile of major roads. This is readily accomplished in both vector and grid-cell systems.

The third situation, in figure 8*C*, is one in which the line does not represent basic data but is simply a boundary between two regions or classes. In a vector system, the line would be labeled with the classes on either side. In a grid-cell system, the line would not exist of itself but would be represented solely by the juxtaposition of the two classes.

The last situation, shown in figure 8*D*, does not involve analysis of the data, but only display. Nevertheless, such a representation can be important in showing the magnitude of an attribute of a line that changes in a manner proportional to some function of the length of the line. It could, for example, represent the increase in mean annual flow of a river in a downstream direction. The author does not know of an algorithm for accomplishing this display.

How does a user frame questions so that they can be answered by a geographic information system? A user with a need for information on a specific area may use a geographic information system to query a geographic data base. In a generic sense, the user may ask one of two questions: (1) For a given area of land, what are its attributes? and (2) For a given use of land, which areas have the proper attributes? The first question is readily answered by displaying, in map form, the various data planes in a geographic data base to allow review of the attributes of the land. The information gained by this approach relies heavily on the way in which the data are stored and classified in the system. A simple geographic data base for an area in the Western United States might, for example, contain the following data planes:

1. Elevation of the land surface, in 200-foot intervals.
2. The slope of the land surface, in intervals of 5 percent.
3. A vegetation map showing barren land, desert shrub, hardwoods, and conifers.
4. A land-use map showing rangeland, cropland, and urban land.

The user can display all of these in turn and develop a perception of the type and condition of the land: it is generally flat or steep, it is predominantly rural or urban, it is largely desert or wooded, and so on. The development of such a perception is important to the user because it allows him to compare, in general terms, many different areas.

The second type of question is more difficult to answer. The answer depends heavily on the classification of information in the system and also very heavily on the way in which the user asks his questions. The system contains the attributes of the land, classified in a certain manner, but the user must understand the manner in which the data are classified in order to frame a specific question. In the example given above, it would be of no use for the user to ask the question, "Where are the areas that are suitable for diversion of water for irrigation?" because the system as described contains no information on water. Although this may seem at first glance to be not only obvious but trivial, it is not. The creator of the geographic data base creates it for a purpose (either explicitly or implicitly) by deciding which information to put in and which information to leave out. He also decides how to classify or group the information that is in each data plane. This process automatically limits the questions that can be asked. It is obviously impossible to put in all the possible data on an area—and it is equally impossible to anticipate all the questions about the area that someone may wish to ask. The user, therefore, must be able to frame his question within the boundaries and parameters of the data and the means of manipulating it. A logical question to be asked, in the above example, could be, "What areas are the best sites for a vacation home in the area?" The question may be within the bounds of the data, but it must be made more specific to cope with the parameters of the data. The reframed question might well be, "Where are the places in this area where the elevation is above 8,000 feet (for coolness and comfort), where the slope is less than 5 percent (for ease of home construction), where there are conifers (because I like such a setting), and where there is little development (more than 5 miles from the nearest farm or town)?" With these bounds that relate to the stored attributes, the question can be answered. The answer will be a map showing a limited number of places that meet the criteria. The user can then explore these areas to pick a specific site by gathering further information to answer such questions as, "Is there a good scenic view from the site?" or "Is the site for sale?"—pertinent questions whose answers are needed for final selection, but ones that cannot be answered by the data in the particular geographic data base.

The use of logical relations for mapping with a digital geographic information system can be demonstrated by starting with the question, "In a given river basin, where are the liquid waste disposal sites that are in the same grid cell as water bodies or are adjacent to grid cells containing water bodies?" This exercise, though intuitively simple, will be described in some detail to show the logic that must be used.

For the river basin, we initially have three digital maps: (1) the outline or area of the drainage basin, (2) the water bodies (lakes and streams) in the basin, and (3) the location of waste sites in the basin. We thus have 3 two-valued sets: (1) basin or not basin $(B \cdot \sim B)$, (2) water body or not water body $(W \cdot \sim W)$, and (3) disposal site or not disposal site $(D \cdot \sim D)$. In this example, the grid-cell size is 1 kilometer square. A grid cell is labeled a "disposal site" if a site occurs anywhere within the grid cell; the same is true for water bodies.

The water-bodies set is treated with a proximity mapping program to map all grid cells that are adjacent to water bodies. This creates a fourth data set $(\hat{W} \cdot \sim \hat{W})$. We can now consider which sets to map and which intersections (logical products) of the sets to map to show their relationships. The data sets are now

1. The area of the basin (B)
2. (a) Water bodies in the basin $(W \times B)$
   (b) Grid cells adjacent to water bodies in the basin $(\hat{W} \times B)$
   (c) Grid cells far from water bodies in the basin $\sim(\hat{W} \times B)$
3. Waste disposal sites in the basin $(D \times B)$

The intersections (logical products) of the data sets are

1. (a) Grid cell with water body and waste site $(W \times B) \times (D \times B)$
   (b) Grid cell with water body and no waste site $(W \times B) \times (\sim D \times B)$
2. (a) Grid cell adjacent to water body with waste site $(\hat{W} \times B) \times (D \times B)$
   (b) Grid cell adjacent to water body with no waste site $(\hat{W} \times B) \times (\sim D \times B)$
3. (a) Grid cell far from water body with waste site $\sim(\hat{W} \times B) \times (D \times B)$
   (b) Grid cell far from water body with no waste site $\sim(\hat{W} \times B) \times (\sim D \times B)$

To answer our original question, the sites are

$$(W \times B) \times (D \times B) + (\hat{W} \times B) \times (D \times B) = (W + \hat{W}) \times D \times B$$

The data sets and their intersections can be represented by a Venn diagram (fig. 9) which shows the logical placement of each data set and the sets' intersections. Three sets and three intersections of sets result as usable products. The Venn diagram aids consideration of all the sets, places them in their correct relations, and provides a graphic means of considering the sets and intersections to be portrayed on a map.

Figure 10 shows the basic data sets in map form in a two-valued or binary form. In each map, the set is black and the negative of the set within the river basin is gray ($a \cdot \sim a$).

The intersections of the sets are also shown in figure 10. Assume that a basic data set shows all the waste sites in a digital form, with the waste sites having values of 1 and the background having a value of 0. A second set has all the water bodies with a value of 1 and the background a value of 0. An "and" (equivalent to the intersect operation) program compares the two data sets and produces an output map in which a grid cell is given a value of 1 only if that cell has a value of 1 in both data sets. This creates a map showing all waste disposal sites within grid cells containing water bodies ($W \times D$). A similar method is used for the other relations of waste sites and grid cells.

The total combination of the basic data sets and their relations is most clearly shown in color. Figure 11 shows the sets and their relations in gray shades. The logical relations used to create the final map are not evident at first glance, and thus the creator of the map should ensure, by means of the Venn diagram, that the logical relations are correct and that all sets and intersections of interest are included and correctly portrayed. By expressing all operations in logical notation, that correctness of the result can be demonstrated mathematically.

Other logical relations may be analyzed and portrayed in a similar manner if so desired.

Hofstadter (1979, p. 297) states, in referring to the flexibility or rigidity of computers:

One of the major goals of the drive to higher levels has always been to make as natural as possible the task of communicating to the computer what you want it to do. . . . When you stop to think what most people use computers for, you realize that it is to carry out very definite and precise tasks which are too complex for people to do. If the computer is to be reliable, then it is necessary that it should understand, without the slightest chance of ambiguity, what it is supposed to do.

The user of a geographic information data base and system usually does not understand, and probably does not need to understand, all of the operations that take place within the computer system when it is trying to formulate an answer to his question. But it is important for him to understand how the data is categorized or classified and whether it relates to the question he is trying to have answered. In turn, it is the task of the applications programmer to have a basic understanding of the types of questions the user may ask and to be able to write the programs that can provide the answers.

Thus, there is a translation problem between the human language in which the questions are framed and the machine language in which the questions are answered. It is in this translation area that geographic information systems will have their greatest problems—but it is also the area where there are the greatest opportunities for making the systems truly usable, reliable, and understandable by the users. It is useful to be able to express, in symbolic language, what logical steps need to be followed to process the data to achieve a desired map. If we cannot clearly express ourselves, how can we expect a computer to figure out what we mean?

Two types of display may result from the combination of data by a geographic information system. The first is the result of the operation of some principle upon the data. The principle may be a mathematical formula expressed as an algorithm and program, and its results may not be intu-

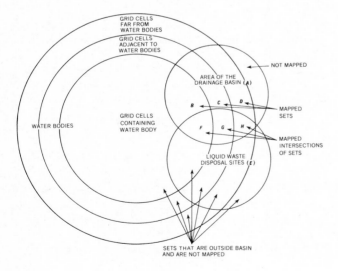

FIGURE 9.—Venn diagram of mapped sets and intersections of sets for water bodies and disposal sites in a river basin. Letters refer to the mapped sets and intersections in figure 10.

itively predictable in magnitude and type from a simple inspection of the (original revised) data. The second is a simple combination of two or more types of data which, when displayed in a pictorial or cartographic form, convey an impression to the viewer of the relations of the data and which is designed to create a psychological reaction to the combined data. The former type of display is "scientific" in that it attempts to discuss the relations among data elements through a process of mathematical logic (modeling) which is not a result of the preconception of the investigator but which is solely the result of the testing of a hypothesis. The display of the results of a model is done in such a way that the relationships are clear and explicit and are not influenced by any constraints outside the model itself. The latter type of display is "propagandistic" in that it portrays relations among data elements in such a way as to influence the viewer to perceive them in the same way as the creator. An example of this is the portrayal on a map of areas where ground water may be susceptible to pollution from industrial waste. Such a map may portray in red the areas that are highly susceptible, in yellow the areas in which caution should be used in waste disposal, and in green the areas where waste may be safely stored. The impression created by such colors on a map (often called stoplight maps) reflect the conclusions of the creator of the map and may be based on less than totally accurate or less than complete data. The scientific portrayal of the data elements and their relationships would perhaps result in a map showing three categories: (1) areas where the water table is deep and the soils are of very low permeability, (2) areas where either the water table is deep or the soils are of low permeability, and (3) areas where the water table is shallow and the soils are highly permeable (note the necessary use of fuzzy terminology.) This would allow the viewer to draw his own conclusions on the basis of the data or to explicitly or implicitly add his own knowledge to the available data to come to a conclusion as to safety. A fairer way to present the information is to show both types of maps so that the user may see both the technical results and the conclusions.

Either of the two methods described above can be legitimately used, with the warning that the maps showing conclusions must be originally based on sound data and on the proper method of relating the data elements. The term "propagandistic" used above is not meant in a pejorative sense but is used to highlight the fact that opinion is added to facts before they are presented (and opinions are always one-sided).

Geographic information systems can be used in both ways, and it is imperative to understand which method is being used in any particular case. The results of analysis of several spatially distributed data elements is valid only insofar as the model and the logic relating them are valid and complete. It is necessary to take into account not only the mathematical and statistical relationships among the elements, but their spatial relationships as well. Co-occurrence of critical data elements in the spatial domain must be of significance or there is no good reason to display them on a map or geographic information system format.

THE PROBLEM OF HOLISM AND REDUCTIONISM

Geographic information systems tend toward the ultimate in reductionism (generally parametric), while integrated mapping systems (Robinove, 1979) tend toward holism. This is a fundamental conflict both on a philosophical basis and on the basis of identifying and solving real problems. Reductionism considers the universe to be composed of separate parts or entities which, in various combinations, make up the whole. Holism considers the universe to be a whole rather than simply the sum of its parts. It is a natural human tendency to separate a whole into its parts, to categorize and classify, to draw boundaries between parts, and to define classes on the basis of rigidly defined boundaries. Boundaries so defined may be useful for some purposes, but they may badly confuse the accomplishment of other purposes. Hofstadter (1979, p. 251) states, "As soon as you perceive an object, you draw a line between it and the rest of the world; you divide the world, artificially, into parts***."

Much of science depends on such division and classification to make sense of the universe. The Earth is divided into the lithosphere, the atmosphere, the hydrosphere, and the overlapping biosphere. The lithosphere is divided, in turn, into various types of rocks on the basis of genetic or mineralogical criteria, and the minerals are themselves divided on the basis of physical or chemical properties. A "landscape" (using the term in the broadest sense) is composed of elements of the lithosphere, atmosphere, hydrosphere, and biosphere. Problems of the use of the landscape by man involve classification of the landscape in such a manner that the classification

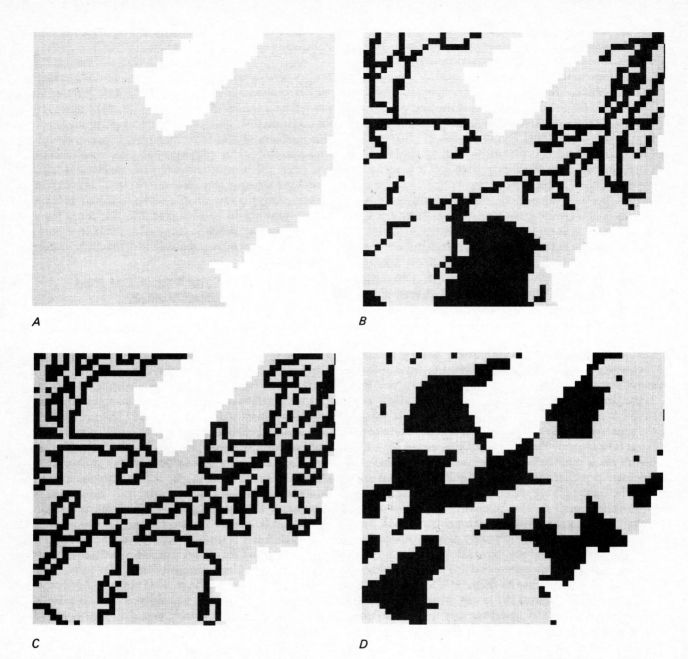

A

B

C

D

FIGURE 10.—Sets and intersections of sets of water and waste-disposal sites in a river basin. In each illustration, the area of the river basin is grey, the area outside the river basin is white, and the set or intersection of interest is black. Relations are shown graphically in the Venn diagram in figure 9. *Sets*: (*A*) River basin, B. (*B*) Water bodies in the river basin, W×B. (*C*) Grid cells immediately adjacent to water bodies in the river basin, Ŵ×B. (*D*) Grid cells far from water bodies in the river basin, ∼(Ŵ+W)×B. The sum of *B*, *C*, and *D* is *A*, the total area of the river basin, (W×B)+(Ŵ×B)+ ∼(W×W)B=B.

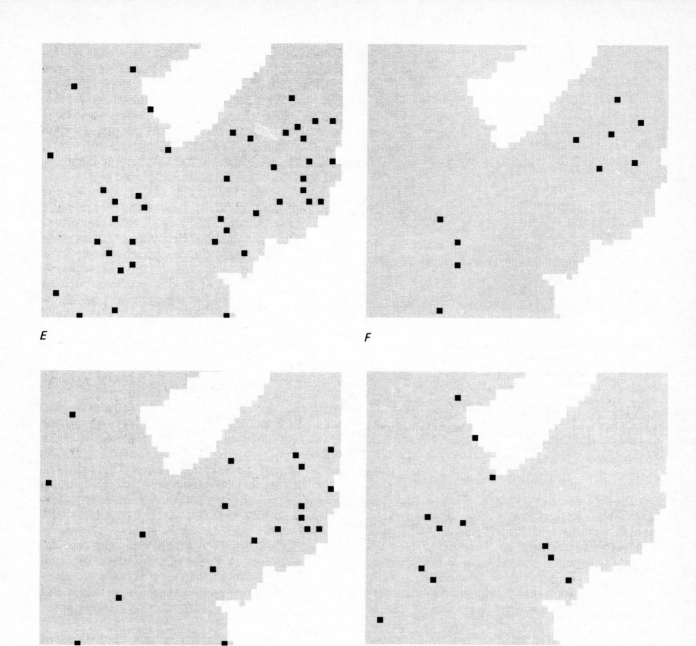

E

F

G

H

FIGURE 10.—Continued. (*E*) All waste disposal sites in the river basin, D×B. *Intersections of sets*: (*F*) Waste disposal sites within grid cells containing water bodies, (D×B)×(W×B) (intersection of *B* and *E*). (*G*) Waste disposal sites in grid cells immediately adjacent to water bodies, (D×B)×(W×B) (intersection of *C* and *E*). (*H*) Waste disposal sites in grid cells far from water bodies, (D×B)× ~(Ŵ+W)×B (intersection of *D* and *E*).

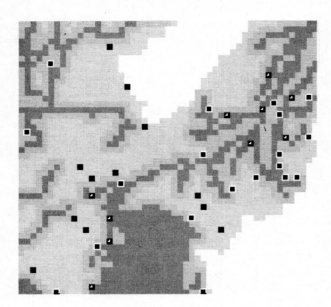

FIGURE 11.—Combination of individual sets and intersections of sets. White—areas outside the drainage basin, ~B. Gray shades—areas within the drainage basin, B: Dark Gray—grid cells containing water, W×B; Medium Gray—grid cells adjacent to water, $\hat{W}$×B; Light Gray—grid cells far from water, ~($\hat{W}$+W)×B. Black cell with white symbol—waste disposal site in grid cell with water, D×W×B. Black cell with white border—waste disposal site in grid cell adjacent to water, D×$\hat{W}$×B. Black cell—waste disposal site far from water, D×~($\hat{W}$+W)×B above w.

is applicable to the problem. Such a problem might be the conversion of a natural grassland to irrigated farming with the attendant buildup of an economic infrastructure. Considered in a reductionistic manner, the solution to the problem must involve the individual facets such as the grass, the soil, the water resources, the engineering properties of the site, transportation, and so forth. This is a prime use of geographic information systems, in which all of the known parametric attributes of the site are defined, measured, and mapped and the geographic information system relates them to the problem and the solution. Considered in a holistic manner, the relations among the attributes of the site become the primary focus and, in addition, the relation of the site to the surrounding area is considered.

The operation of a geographic information system depends on two factors, (1) the availability in the proper form and format of all of the data that are needed and (2) the availability of the relational or process model that operates on the data to produce a solution to a problem. At the present state of the art, it is possible to collect and place in

the proper form the data that are required in a reductionistic manner, which means that the information that is considered to be of significance to the problem is classified into categories that are relevant. The data are manipulated by mathematical, statistical, or Boolean logical operations to produce problem solutions. At the present time the solutions are basically reductionistic.

Holistic solutions are possible only if the original data are collected and represented in a holistic fashion and the algorithms that operate on the data represent a holistic view of the problem and its solution. An example may serve to clarify the difference between the two views. In an area where animal grazing is the principal use of the land, it is necessary to manage the land to maximize the production of animal products without degrading the carrying capacity of the land below a certain sustainable level. A reductionist method of mapping the carrying capacity of the land involves classifying each of the factors of geomorphic features, soils, and vegetation and modeling these to determine the carrying capacity. A holistic method involves integrated mapping of the complex of geomorphic features, soils, and vegetation, and from this deriving the carrying capacity of the individual complexes. A major tenet of the holistic integrated mapping approach is that each parcel of land is considered unique. This is in contrast to the reductionistic view, which sets up classification schemes and then maps each parcel of land in such a way that it must fit into a niche in a selected classification. In essence, the reductionistic view makes it difficult to map and describe ecotones, or areas of transitions between classes, while the holistic view considers transitions to be more common than pure classes and affords them an easily recognizable and mappable status.

Geographic information system algorithms at present are based almost completely on Boolean logical operations. They are capable of theoretically simple operations that characterize parcels of land in terms of the relations of their reductionistic attributes. More sophisticated models can, of course, be applied to the original data to derive secondary or tertiary attributes, and these later attributes can in turn be manipulated by geographic information systems algorithms. Nevertheless, the ability to model processes with the reductionistic methods of geographic information systems is in its infancy, and the links between modeling in various fields and the geographic information systems are far from being well defined. It should be a goal of research in these fields to achieve a holistic view that will allow the modeling and relational techniques to cope with the complexities of nature and man's influence on it in a complete problem-solving manner.

CONCLUSIONS

Digital geographic information systems operate on logical principles. Those principles should be understood by the people who enter data into systems, by the programmers who design the operations, and by the users who expect answers to problems from the systems. Users must be capable of phrasing their questions in forms that are amenable to the data and the operations.

Many of the rather complicated operations that are performed on data planes in a geographic information system can be reduced to logical primitive operations (illustrated by Venn diagrams and expressed in symbolic notation) that are more readily understood by users than the algorithms and programs that are actually used in the systems. It is hoped that this understanding of those logical operations will aid users in making the most effective use of digital geographic information systems.

SELECTED REFERENCES

Calkins, H.W., and Tomlinson, R.F., 1977, Geographic information systems, methods and equipment for land use planning 1977: Ottawa, Canada, International Geographical Union, 394 p.

Carnap, Rudolph, 1958, Introduction to symbolic logic and its applications, New York, Dover Press.

Eco, Umberto, 1984, The name of the rose: New York, Warner Books, 640 p.

Hofstadter, Douglas, 1979, Godel, Escher, Bach—An eternal golden braid: New York, Basic Books, 777 p.

Jupp, D.L.B., and Mayo, K.K., 1982, The use of residual images in Landsat image analysis. Photogrammetric Engineering and Remote Sensing, v. 48, no. 4, p. 595-604.

Luckey, R.R., and Ferrigno, C.F., 1982, A data-management system for areal interpretive data for the High Plains in parts of Colorado, Kansas, Nebraska, New Mexico, Oklahoma, South Dakota, Texas, and Wyoming. U.S. Geological Survey Water-Resources Investigations 82–4072, 112 p.

Mandelbrot, Benoit, 1977, Fractals: Form, chance, and dimension, New York, W.H. Freeman Co., 297 p.

McHarg, I.L., 1969, Design with nature: New York, Natural History Press, 198 p.

Negoita, C.V., 1981, Fuzzy systems. London, Abacus Press, 111 p.

Nystuen, J.D., 1968, Identification of some fundamental practical concepts, in Berry, J.L.B., and Marble, D.F., Spatial analysis—Reader in statistical geography. Englewood Cliffs, N.J., Prentice-Hall, p. 35-41.

Robinove, C.J., 1979, Integrated terrain mapping with digital Landsat images in Queensland, Australia. U.S. Geological Survey Professional Paper 1102, 39 p.

———1981, The logic of multispectral classification and mapping of land. Remote Sensing of Environment, v. 11, p. 231-244.

Varnes, D.J., 1974, The logic of geological maps with reference to their interpretation and use for engineering purposes. U.S. Geological Survey Professional Paper 837, 48 p.

Werkmeister, W.H., 1949, An introduction to critical thinking: Lincoln, Nebr., Johnson Publishing Co., 663 p.

Zadeh, L.A., 1965, Fuzzy sets and systems. Symposium on System Theory, Polytechnic Institute of Brooklyn, 1964, Proceedings, p. 29-39.

Fundamental operations in computer-assisted map analysis

JOSEPH K. BERRY†

Yale University, School of Forestry and Environmental Studies,
205 Prospect Street, New Haven, Connecticut 06511, U.S.A.

Abstract. Geographical information systems (GIS) provide capabilities for the mapping, management and analysis of cartographic information. Unlike most other disciplines, GIS technology was born from specialized applications. A comprehensive theory relating the various techniques used in these applications is only now emerging. By organizing the set of analytic methods into a mathematical structure, a generalized framework for cartographic modelling is developed. Within this framework, users logically order primitive operators on map variables in a manner analogous to traditional algebra and statistics. This paper describes the fundamental classes of operations used in computer-assisted map analysis. Several of the procedures are demonstrated using a fourth-generation computer language for personal computers.

1. Introduction

Information related to the spatial characteristics of resources has been difficult to incorporate in decisions on land planning and management. Manual techniques of map analysis are both tedious and analytically limiting. Computer-assisted geographical information systems (GIS), on the other hand, hold promise in providing capabilities clearly needed for effective decisions. This need has caused the developing science of map analysis to be somewhat prematurely cast into operational contexts. The result has been a growing number of special-purpose, expensive systems designed for unique applications. However, recent efforts have attempted to recognize the similarities in processing among such systems and to develop a more generalized set of capabilities and a theoretical framework.

Most GIS include capabilities in processing that relate to the encoding, storage, processing and display of spatial data. This paper describes a series of techniques that relate to the analysis of mapped data and to the needs of natural resource and environmental planning in particular. The fundamental operations of map processing discussed are, however, common to a broad range of applications.

By organizing primitive operations in a logical manner, a generalized approach to cartographic modelling can be developed (Berry 1986 a, Tomlin 1983 a, b, Tomlin and Berry 1979). This fundamental approach can be conceptualized as a 'map algebra' in which entire maps are treated as variables. In this context, primitive operations of map analysis can be seen as analogous to traditional mathematical operations. The sequencing of map operations is similar to the algebraic solution of equations to find unknowns. In this case, however, the unknowns represent entire maps.

This set of independent operations may be assembled into a general-purpose package for map analysis similar to the many non-spatial data-base, statistical and

† Dr. Berry is an Associate Professor and Associate Dean at Yale University, School of Forestry and Environmental Studies.

© 1987, Taylor & Francis, Ltd., *International Journal of Geographical Information Systems*, v. 1, n. 2, p. 119–136; reprinted by permission.

spreadsheet packages. The Professional Map Analysis Package (pMAP) (SIS 1986), used for the demonstrations in this paper, is an example of such a 'toolbox' implemented for use on personal computers.

2. Data and processing structures

In order to use primitive operations in a modelling context, two fundamental conditions must be met: (1) a common data structure, and (2) a flexible processing structure. The variety of mappable characteristics associated with any given geographical location may be organized as a series of spatially-registered computer-compatible maps, or 'overlays'. An overlay is simply a special form of a geographical map with each cartographic location having one and only one thematic attribute (i.e. mutually exclusive in space). By contrast, a composite map, such as a conventional U.S. Geological Survey topographic sheet, has each location characterized simultaneously by several themes, such as its elevation, water features, vegetation and political designation. In overlay format, each of these individual characteristics would be stored as separate data planes, in the same manner as colour separation negatives used in printing. For convenience, the following discussions use the terms 'overlay' and 'map' interchangeably.

In terms of data structure, each overlay is comprised of a title, certain descriptive parameters and a set of categories, termed 'regions'. Formally stated, a region is simply one of the thematic designations on an overlay. An overlay of land cover entitled COVERTYPE, for example, might include regions associated with open water, meadow and forest. Each region can be represented by a name (i.e. label) and a numerical value. The structure as described so far, however, does not account for locational characteristics. The handling of locational information is not only what most distinguishes processing of geographical information from other types of automated data processing; it is also what most clearly distinguishes one GIS from another.

There are two basic approaches in representing locational information: (1) polygon (or line segment), and (2) gridded (or raster). The former approach stores information about the boundaries between regions, whereas the latter stores information on the interiors of regions. While these differences are significant in terms of strategies for implementation and may vary considerably in terms of geographical precision, they need not affect the definition of a set of fundamental analytical techniques. In the light of its conceptual simplicity, the gridded structure is best suited to the description of primitive techniques for map processing and is used for discussion in this article.

The commonly-used gridded data structure is based on the condition that all spatial locations are defined with respect to a cartographic grid of numbered rows and columns. As such, the smallest addressable unit of space corresponds to a square parcel of land, or what is formally termed a 'point'. Spatial patterns are represented by assigning all points within a particular region according to the numerical value of the region. Also, each point can be addressed as part of a 'neighbourhood' of surrounding values. Figure 1 illustrates the relationship between values, points, regions and neighbourhoods for gridded data. In a polygonal data structure, the smallest addressable unit is the polygon. For actual point features, these polygons have neither length nor width. Linear features are polygons without width. Regions on an overlay are formed by sets of polygons of the same thematic value, and neighbourhoods are identified as sets of surrounding polygons.

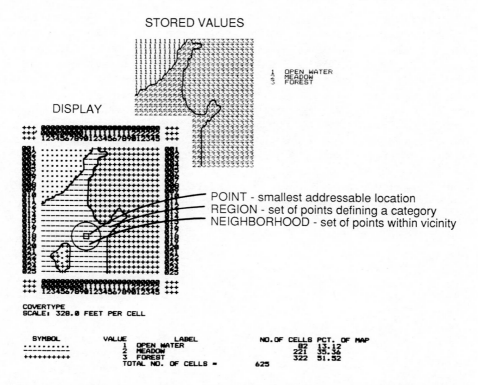

STORED VALUES

DISPLAY

1 OPEN WATER
2 MEADOW
3 FOREST

POINT - smallest addressable location
REGION - set of points defining a category
NEIGHBORHOOD - set of points within vicinity

COVERTYPE
SCALE: 328.0 FEET PER CELL

SYMBOL	VALUE	LABEL	NO. OF CELLS	PCT. OF MAP
....	1	OPEN WATER	82	13.12
----------	2	MEADOW	221	35.36
++++++++++	3	FOREST	322	51.52
		TOTAL NO. OF CELLS =	625	

Figure 1. Gridded data structure. Each overlay characterizes a single theme and is comprised of regions identified by labels and represented by numerical values. These values are then associated with grid cells, formally termed points, explicitly or implicitly identified by row and column coordinates, and displayed as a graphic symbol. Sets of points having the same value identify regions (e.g. meadow category outlined). Sets of points within the same vicinity identify neighbourhoods.

If primitive operations are to be combined flexibly, a processing structure must be used that accepts input and generates output in the same data format. This may be accomplished by requiring that each analytic operation involves

retrieval of one or more overlays from the data file;
manipulation of those data;
creation of a new overlay whose categories are represented by thematic values
defined as a result of that manipulation;
storage of that new overlay for subsequent processing.

The cyclical nature of this processing structure is analogous to the 'evaluation of nested parentheticals' in traditional algebra. The logical sequencing of primitive operations on a set of maps forms a cartographic model of specified application. As with traditional algebra, fundamental techniques involving several primitive operations can be identified (e.g. a 'travel-time' map) that are applicable to numerous situations. The use of primitive analytical operations in a generalized modelling context accommodates a variety of analyses in a common, flexible and intuitive manner. It also provides a framework for instruction in the principles of computer-assisted map analysis that stimulates the development of new techniques and applications (Berry 1986 b, c).

The creation of a new overlay is a function of the values defining the input maps (i.e. variables) and of the transformation. The result is a map of new values having the same spatial registration as the input maps. The table outlines the fundamental classes of operations in map analysis. The first distinction, based on map components, identifies whether the new value assigned to each location is a function of the values associated with the point itself, its region or its neighbourhood. This classification notes how the computer obtains values for processing. Point processors are spatially myopic and consider each geographical unit independently. When considering several overlays, the process can be conceptualized as 'vertically spearing' a series of values from a stack of spatially-registered maps. Regional processors, on the other hand, associate each location with a set of other locations having a similar characteristic. This process can be conceptualized as forming 'cookie cutter' templates (i.e. regions) which are superimposed on other maps. The values occurring within the boundaries of each

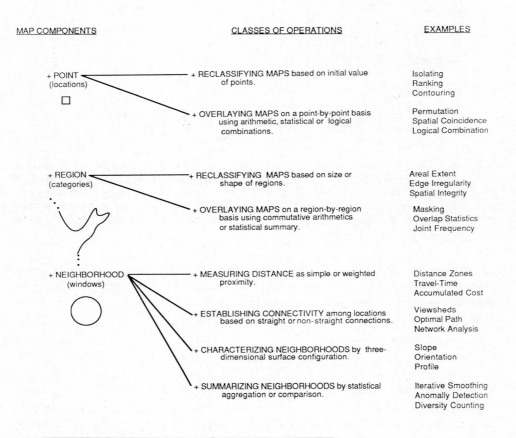

MAP COMPONENTS	CLASSES OF OPERATIONS	EXAMPLES
+ POINT (locations)	+ RECLASSIFYING MAPS based on initial value of points.	Isolating Ranking Contouring
	+ OVERLAYING MAPS on a point-by-point basis using arithmetic, statistical or logical combinations.	Permutation Spatial Coincidence Logical Combination
+ REGION (categories)	+ RECLASSIFYING MAPS based on size or shape of regions.	Areal Extent Edge Irregularity Spatial Integrity
	+ OVERLAYING MAPS on a region-by-region basis using commutative arithmetics or statistical summary.	Masking Overlap Statistics Joint Frequency
+ NEIGHBORHOOD (windows)	+ MEASURING DISTANCE as simple or weighted proximity.	Distance Zones Travel-Time Accumulated Cost
	+ ESTABLISHING CONNECTIVITY among locations based on straight or non-straight connections.	Viewsheds Optimal Path Network Analysis
	+ CHARACTERIZING NEIGHBORHOODS by three-dimensional surface configuration.	Slope Orientation Profile
	+ SUMMARIZING NEIGHBORHOODS by statistical aggregation or comparison.	Iterative Smoothing Anomaly Detection Diversity Counting

"Classes of Analytic Operations"-- Each primitive operation may be regarded as an independent tool limited only by the general thematic and spatial characteristics of the data. From this point of view, four classes of fundamental map analysis operations may be identified:

* Reclassifying Maps
* Overlaying Maps
* Measuring Distance and Connectivity
* Characterizing Neighborhoods

template become available for processing. Note that these templates are not necessarily contiguous clumps, but can be a collection of groupings of the same theme throughout the map (e.g. the forest category in figure 1). Neighbourhood processors, like regional ones, are dependent upon the spatial arrangement of values. In this instance, however, values for processing are identified by specified distance and directional proximity. These 'roving windows' are centred about each location and the surrounding values become available for processing.

The next distinction is based on the processing transformation. From this perspective, four classes of primitive operations may be identified, viz., those which

reclassify map categories;
overlay maps on a point-by-point or region-wide basis;
measure simple or weighted distance and connectivity;
characterize cartographic neighbourhoods.

Reclassification operations merely repackage existing information on a single map. Overlay operations, on the other hand, involve two or more maps and result in delineations of new boundaries. Distance and connectivity operations generate entirely new information by characterizing the juxtapositioning of features. Neighbourhood operations summarize the conditions occurring in the general vicinity of a location. The reclassifying and overlaying operations based on point processing are the backbone of current GIS applications, allowing rapid updating and examination of mapped data. However, other than the significant advantage of speed and ability to handle tremendous volumes of data, these capabilities are similar to those of techniques of manual processing. The operations based on regional and neighbourhood assessment identify more advanced analytic capabilities.

3. Reclassifying maps

The first and in many ways the most fundamental class of analytical operations involves the reclassification of map categories. In each operation a new map is created by assigning thematic values to the categories of an existing overlay. These values may be assigned as a function of the initial value, its position, contiguity, size or shape of the spatial configuration of the individual categories. Each of the reclassification operations involves the simple repackaging of information on a single overlay and results in no delineation of new boundaries. Such operations can be thought of as the purposeful 'recolouring' of maps.

Figure 2 shows the results of simple reclassifying a map as a function of its initial thematic values. For display, a unique symbol is associated with each value. The COVERTYPE map has categories of open water, meadow and forest. These features are stored as thematic values 1, 2 and 3, respectively, and displayed as separate graphic patterns. Colour codes could be used in place of the patterns for more elaborate graphic presentation. A binary map that isolates the meadow can be created simply by assigning 0 to the open and forested areas, and displaying the graphic symbol ' ' (blank) wherever this value occurs.

A similar reclassification operation might involve the ranking or weighting of qualitative map categories to generate a new map with quantitative values. A map of soil types, for example, might be assigned values that indicate the relative suitability of each type for residential development. Quantitative values may also be reclassified to yield new quantitative values. This might simply involve a specified reordering of map categories (e.g. generating a map of levels of suitability for plant growth from a map of

85

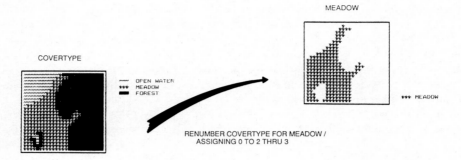

Figure 2. Reclassifying maps. Reclassification of map categories can be based on initial thematic value, as in this example. The cover types of open water and forest are renumbered to the value zero which is displayed as a blank. The resulting map isolates the meadow region.

soil moisture content). Alternatively, it could involve the application of a generalized reclassifying function, such as 'level slicing', which splits a continuous range of values for each map category into discrete intervals (e.g. derivation of a contour map from a map of values of topographic elevation). Other quantitative reclassifying functions include a variety of arithmetic operations involving values for each category and a specified or computed constant. Among these operations are addition, subtraction, multiplication, division, exponentiation, maximization, minimization, normalization and other scalar mathematical and statistical operators. For example, a map of topographic elevation expressed in feet may be converted to one expressed in metres by multiplying each map value by the appropriate conversion factor of 3·28083 feet per metre.

Reclassification operations can also relate to locational, as well as purely thematic, attributes associated with a map. One such characteristic is position. An overlay category represented by a single 'point' location, for example, might be reclassified according to its latitude and longitude. Similarly, a line segment or areal feature could be reassigned values indicating its centre or general orientation. A related operation, termed 'parcelling', characterizes the contiguity of categories. This procedure identifies individual 'clumps' of one or more points having the same numerical value and that are spatially contiguous (e.g. generation of a map identifying each lake as a unique value from a generalized map of water on which all lakes are represented as a single category).

Another locational characteristic is size. In the case of map categories associated with linear features or point locations, overall length or number of points might be used as the basis for reclassifying those categories. Similarly, an overlay category associated with a planar area might be reclassified according to its total acreage or the length of its perimeter. For example, an overlay of surface water might be reassigned values to indicate the areal extent of individual lakes or the length of stream channels. The same sort of technique might also be used to deal with volume. Given a map of depth to bottom for a group of lakes, for example, each lake might be assigned a value indicating total volume of water based on the areal extent of each depth category.

In addition to the value, position, contiguity and size of features, categories may be reclassified on the basis of shape. Categories represented by point locations have measurable 'shapes' in so far as the set of points implies linear or areal forms (i.e. just as stars imply constellations). Characteristics of shape associated with linear forms identify the patterns formed by multiple line segments (e.g. dendritic stream patterns). The primary characteristics of shape associated with areal forms include topological genus, convexity of boundaries and nature of edge. Topological genus relates to the 'spatial integrity' of an area. A category that is broken into numerous 'fragments' and/or contains several interior 'holes' is said to have less spatial integrity than one without such violations. The topological genus can be summarized as the 'Euler number' which is computed as the number of holes within a feature less one short of the number of fragments which make up the entire feature. An Euler number of zero indicates features that are spatially balanced, whereas larger negative or positive numbers indicate less spatial integrity.

The other characteristics of shape, i.e. convexity and edge, relate to the configuration of the boundaries of areal features. Convexity is the measure of the extent to which an area is enclosed by its background, relative to the extent to which the area encloses this background. The 'convexity index' for a feature is computed by the ratio of its perimeter to its area. The most regular configuration is that of a circle which is totally convex, and therefore not enclosed by the background at any point along its boundary. Comparison of a feature's computed convexity with that of a circle of the same area results in a standard measure of the regularity of boundaries. The nature of the boundary at each edge point can be used for a detailed description of its configuration. At some locations the boundary might be an entirely concave intrusion, whereas other boundary points might be at entirely convex protrusions. Depending on the 'degree of edgeness' each point can be assigned a value indicating the actual convexity of boundary at that location.

This explicit use of cartographic shape as an analytic parameter is unfamiliar to most users. However, a non-quantitative consideration of shape is implicit in any visual assessment of mapped data. Particularly promising is the potential for applying techniques of quantitative shape analysis in the classification of digital images and the modelling of wildlife habitats. A map of forest stands, for example, might be reclassified so that each stand is characterized according to the relative amount of forest edge with respect to total acreage and the frequency of gaps in the interior of the forest canopy. Those stands with a large proportion of edge and a high frequency of gaps will generally indicate better habitats for many species of wildlife.

4.   Overlaying operations

Operations for overlaying maps begin to relate to spatial coincidence, as well as to the thematic nature of cartographic information. The general class of overlaying operations can be characterized as 'light table gymnastics'. These involve the creation of a new map on which the value assigned to every point or set of points is a function of the independent values associated with that location on two or more existing overlays. In 'location-specific' overlaying, the value assigned is a function of the point-by-point coincidence of the existing overlays. In 'region-wide' compositing, values are assigned to entire thematic regions as a function of the values on other overlays that are associated with the regions. Whereas the first approach conceptually involves the vertical spearing of a set of overlays, the second approach uses one overlay to identify boundaries by which information is extracted from other overlays.

Figure 3 shows an example of location-specific overlaying. Here, maps of COVERTYPE and topographic SLOPE are combined to create a new map identifying the particular cover/slope combination at each location. A specific function used to compute values of categories on the new map from those of existing maps being overlaid may vary according to the nature of the data being processed and the specific use of those data within a modelling context. Most typical of environmental analyses is the manipulation of quantitative values to generate new values which are likewise quantitative in nature. Among these are the basic arithmetic operations such as addition, subtraction, multiplication, division, roots and exponentiation. For example, given maps of assessed land values in 1965 and 1980 respectively, one might generate a map showing the change in land values over that period as follows (expressed in pMAP software syntax):

**COMPUTE 1980MAP MINUS 1965MAP TIMES 100 DIVIDED
BY 1965MAP/FOR PERCENT-CHANGE-MAP**

Functions which relate to simple statistical parameters such as maximum, minimum, median, mode, majority, standard deviation or weighted average may also be applied in this manner. The type of data being manipulated dictates the appropriateness of the mathematical or statistical procedure used. For example, the addition of qualitative maps such as soils and land use would result in mathematically meaningless sums, since their thematic values have no numerical relationship. Other techniques of map overlay include several which may be used to process either quantitative or qualitative data and generate values which may likewise take either form. Among these are masking, comparison, calculation of diversity and permutations of map categories, as depicted in figure 3.

More complex statistical techniques may also be applied in this manner, assuming that the inherent interdependence among spatial observations can be taken into account. This approach treats each map as a variable, each point as a case and each

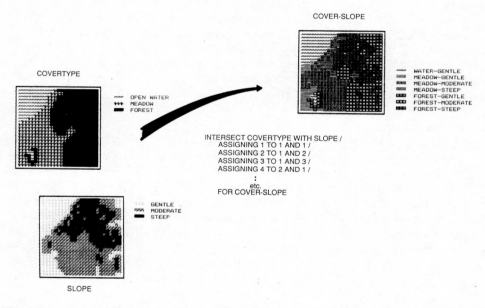

Figure 3.   Point-by-point overlaying. Overlaying of maps is most frequently performed on a location-specific basis. In this example, each map location is assigned a unique value identifying the COVERTYPE and SLOPE conditions occurring at that location.

value as an observation. A predictive statistical model can then be evaluated for each location, resulting in a spatially continuous surface of predicted values. The mapped predictions contain additional information over traditional non-spatial procedures, such as direct consideration of coincidence among regression variables and the ability to locate areas of a given level of prediction.

An entirely different approach to overlaying maps involves region-wide summarization of values. Rather than combining information on a point-by-point basis, this group of operations summarizes the spatial coincidence of entire categories of two or more maps. Figure 4 contains an example of a region-wide overlay operation using the same input maps as those in figure 3. In this example, the categories of the COVER type map are used to define areas over which the coincidental values of the SLOPE map are averaged. The computed values of average slope are then assigned to each of the categories of cover type. Summary statistics which can be used in this way include the total, average, maximum, minimum, median, mode or minority value; the standard deviation, variance or diversity of values; and the correlation, deviation or uniqueness of a particular combination of values. For example, a map indicating the proportion of undeveloped land within each of several counties could be generated by superimposing a map of county boundaries on a map of land use and computing the ratio of undeveloped land to the total land area for each county. Or a map of postal codes could be superimposed over maps of demographic data to determine the average income, average age and dominant ethnic group within each zip code.

As with location-specific overlay techniques, types of data must be consistent with the summary procedure used. Also of concern is the order of data processing.

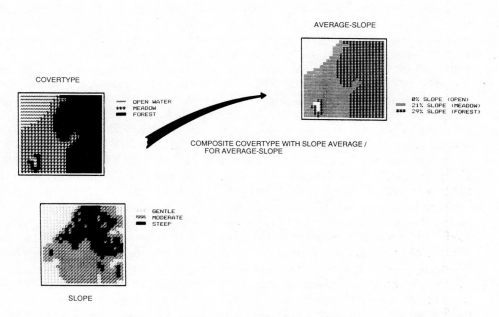

Figure 4.   Region-wide overlaying. These operations summarize the spatial coincidence of regions. In this example, each of the three COVERTYPES (open water, meadow and forest) is assigned a value equal to the average of the SLOPE values occurring within its boundaries.

Operations such as addition and multiplication are independent of the order of processing. Other operations, such as subtraction and division, however, yield different results depending on the order in which a group of numbers is processed. This latter type of operation, termed non-commutative, cannot be used for region-wide summaries.

5. Measuring distance and connectivity

Most GIS contain analytic capabilities for reclassifying and overlaying maps. These operations address the majority of applications that parallel manual techniques of map analysis. However, to integrate spatial considerations more fully into decision making, new techniques are emerging. The concept of distance has been historically associated with the 'shortest straight-line distance between two points'. While this measure is both easily conceptualized and implemented with a ruler, it is frequently insufficient in decision making. A straight line route may indicate the distance 'as the crow flies', but offer little information for a walking or hitchhiking crow or other flightless creature. It is equally important to most travellers to have the measurement of distance expressed in more relevant terms, such as time or cost. The operations concerned with measuring this effective distance are best characterized as 'rubber rulers'.

Any system for the measurement of distance requires two components: a standard unit and a procedure for measurement. The unit used in most computer-oriented systems is the 'grid space' implied by the superimposing of an imaginary grid over an area. A ruler with its uniform markings implies such a grid each time it is laid on a map. The procedure for measuring actual distance from any location to another involves counting the number of intervening grid spaces and multiplying by the map scale. If the grid pattern is fixed the length of the hypotenuse of a right-angled triangle formed by the grid is computed. This concept of point-to-point distance may be expanded to one of proximity. Rather than sequentially computing the distance between pairs of locations, concentric equidistant zones are established around a location or set of locations. This is analogous to the wave pattern generated when a rock is thrown into a still pond. Insert (*a*) in figure 5 is an example of a simple proximity map indicating the shortest, straight-line distance from the ranch to all other locations. A more complex proximity map would be generated if, for example, all locations of houses were considered target locations: in effect, throwing a handful of rocks into the pond. The result would be a map or proximity indicating the shortest straight-line distance to the nearest target area (i.e. house) for each non-target location.

In many applications, however, the shortest route between two locations may not always be a straight line. And even if it is straight, the Euclidean length of that line may not always reflect a meaningful measure of distance. Rather, distance in these applications is best defined in terms of movement expressed as travel-time, cost or energy that may be consumed at rates which vary over time and space. Distance-modifying effects may be expressed cartographically as 'barriers' located within the space in which the distance is being measured (Berry and Tomlin 1982 a), an approach which implies that distance is the result of some sort of movement over that space and through those barriers. Two major types of barriers can be identified by their effects on the implied movement. 'Absolute barriers' are those which completely restrict movement and therefore imply an infinite distance between the points they separate, unless a path around the barrier is available. A river might be regarded as an absolute barrier to a non-swimmer. To a swimmer or a boater, however, the same river might be regarded as a relative rather than an absolute barrier. 'Relative barriers' are those that

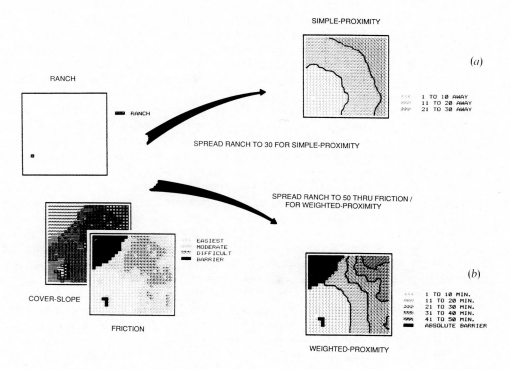

RANCH

■ RANCH

SIMPLE-PROXIMITY

(a)

```
::::   1 TO 10 AWAY
:::::  11 TO 20 AWAY
:::::  21 TO 30 AWAY
```

SPREAD RANCH TO 30 FOR SIMPLE-PROXIMITY

SPREAD RANCH TO 50 THRU FRICTION /
FOR WEIGHTED-PROXIMITY

```
::::   EASIEST
:::::  MODERATE
:::::  DIFFICULT
■■■■   BARRIER
```

COVER-SLOPE

FRICTION

(b)

```
::::   1 TO 10 MIN.
:::::  11 TO 20 MIN.
:::::  21 TO 30 MIN.
:::::  31 TO 40 MIN.
■■■■   41 TO 50 MIN.
■■■■   ABSOLUTE BARRIER
```

WEIGHTED-PROXIMITY

Figure 5.   Measurement of distance. Distance between locations can be determined as simple distance or as a function of the effect of absolute or relative barriers on implied movement. In this example, insert (a) identifies equidistant zones around the ranch. Insert (b) is a map of hiking travel-time from the ranch. It was generated by considering the relative ease of travel through various cover and slope conditions (see figure 4) where flat meadows are the fastest to traverse, steep forested areas are intermediate and flat water is an absolute barrier to travel.

are passable, but only at a cost which may be equated with an increase in physical distance.

Insert (b) of figure 5 shows a map of hiking time around the target location identified by the ranch. The map was generated by relassifying the various cover/slope categories (see figure 3) in terms of their relative ease of travel on foot. In the example, two types of barriers are used. The lake is treated as an absolute barrier, completely restricting hiking. The land areas, on the other hand, represent relative barriers which indicate varied impedance to hiking at each point as a function of the cover/slope conditions occurring at that location. In a similar manner, movement by automobile may be effectively constrained to a network of roads (absolute barriers) of varying speed limits (relative barriers) to generate a map of travel time by the mode. Or, from an even less conventional perspective, weighted distance can be expressed in such terms as accumulated cost of constructing power lines from an existing trunk-line to all other locations in a study area. The cost surface that is developed can be a function of a variety of social and engineering factors, such as visual exposure and adverse topography, expressed as absolute and/or relative barriers.

The ability to move, whether physically or abstractly, may vary as a function of the implied movement as well as of the static conditions at a location. One aspect of movement which may affect the ability of a barrier to restrict that movement is direction. A topographic incline, for example, will generally impede hikers differently depending on whether their movement is uphill, dowhill or across the slope. Another possible modifying factor is accumulation. After hiking a certain distance, 'molehills' tend to become disheartening 'mountains', and movement is more restricted. A third attribute of movement which might dynamically alter the effect of a barrier is momentum, or speed. If an old car is stopped on a steep hill, it may not be able to resume movement, whereas if it were allowed to maintain its momentum (e.g. a green traffic light), it could easily reach the top. Similarly, an impairment to a highway which effectively reduces traffic speeds from, say, 55 to 40 miles per hour, would have little or no effect during a rush hour when traffic is already moving at a much slower speed.

Another distance-related class of operations is concerned with the nature of connectivity among locations on an overlay. Fundamental to understanding these procedures is the conceptualization of an 'accumulation surface'. If the thematic value of a simple map of proximity from a point is used to indicate the third dimension of a surface, a uniform bowl would be formed. The surface configuration for a weighted proximity map would have a similar appearance, but the bowl would be warped with numerous ridges and pinnacles. Also, the nature of the surface is such that it cannot contain saddle points (i.e. false bottoms). This 'bowl-like' topology is characteristic of all accumulation surfaces and can be conceptualized as a football stadium with each successive ring of seats identifying concentric, equidistant haloes. The bowl need not be symmetrical, however, and may form a warped surface responding to varying rates of accumulation. The three-dimensional insert in figure 6 shows the surface configuration of the map of hiking travel time from figure 5. The accumulated distance surface is shown as a perspective plot in which the ranch is the lowest location and all other locations are assigned progressively larger values of the shortest distance (not necessarily straight) to the ranch. When viewed in perspective, this surface resembles a topographic surface with familiar valleys and hills. However, in this case the highlands indicate areas that are effectively farther away from the ranch.

In the case of simple distance, the delineation of paths, or 'connectivity', locates the shortest straight line between two points considering only two dimensions. Another technique traces the steepest downhill path from a location over a complex three-dimensional surface. The steepest downhill path along a topographic surface will indicate the route of surface runoff. For a surface represented by a map of travel time, this technique traces the optimal (e.g. the shortest or quickest) route. Insert (*a*) of figure 6 indicates the optimal path for hiking from a nearby cabin to the ranch as the steepest downhill path over the accumulated hiking-time surface shown in figure 5. If an accumulation cost surface is considered, such as the cost surface for constructing powerlines described above, the minimum cost route will be located. If such a construction to a set of dispersed locations was simultaneously considered, a map of 'optimal path density' could be generated which identifies the number of individual optimal paths passing through each location from the dispersed termini to the trunk-line. Such a map would be valuable in locating major feeder lines (i.e. areas of high optimal path density) radiating from the central trunk-line.

Another connectivity operation determines the narrowness of features. The narrowness at each point within a map feature is defined as the length of the shortest line segment (i.e. chord) which can be constructed through that point to diametrically opposing edges of the feature. The result of this processing is a continuous map of features with lower values indicating relatively narrow locations. For a map of

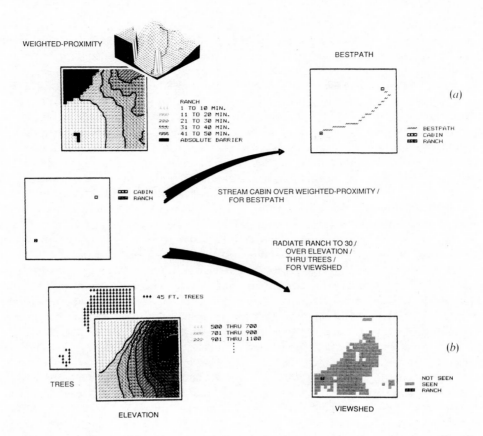

Figure 6. Establishing connectivity. These operations characterize the nature of spatial linkages among locations. Insert (*a*) delineates the shortest (i.e. the least time) hiking route from the cabin to the ranch. The route traces the steepest downhill path along the travel time surface derived in figure 5 (also shown as a three-dimensional plot). Insert (*b*) identifies the viewshed of the ranch. Topographic relief and forest cover act as absolute barriers when establishing visual connectivity.

narrowness of forest stands, low thematic values indicate interior locations with easy access to edges.

The characterization of viewsheds involves establishing intervisibility among locations. Locations forming the viewshed of an area are connected by straight rays in three-dimensional space to the location of the 'viewer' or set of viewers. Topographic relief and surface objects form absolute barriers that preclude connectivity. If multiple viewers are designated, locations within the viewshed may be assigned a value indicating the number or density of visual connections. Insert (*b*) of figure 6 shows a map of the viewshed of the ranch, in which the terrain and the height of the canopy are considered as visual barriers.

6. Characterizing neighbourhoods

The final group of operations includes procedures that create a new map where the value assigned to a location is computed as a function of independent values surrounding that location (i.e. its cartographic neighbourhood) (Berry and Tomlin

1982 b). This general class of operations can be conceptualized as 'roving windows' moving throughout the mapped area. The summary of information within these windows can be based on the configuration of the surface (e.g. slope and aspect) or the mathematical summary of thematic values.

The initial step in characterizing cartographic neighbourhoods is to establish the membership of such a neighbourhood. A neighbourhood is uniquely defined for each target location as the set of points which lie within a specified distance and direction around that location. In most applications, the window has a uniform geometric shape and orientation (e.g. a circle or square). However, as noted in the previous section, the distance may not necessarily be Euclidean or symmetrical, such as a neighbourhood of 'down-wind' locations within quarter of a mile of a smelting plant. Similarly, a neighbourhood of 'the 10-min drive' along a road network could be defined.

The summary of information within a neighbourhood may be based on the relative spatial configuration of values that occur within the window. This is true of the operations which measure topographic characteristics such as slope, aspect or profile from elevation values. One such technique involves the 'least-squares fit' of a plane to adjacent elevation values. This process is similar to fitting a linear regression line to a series of points expressed in two-dimensional space. The inclination of the plane denotes the slope of the terrain and its orientation characterizes the aspect. The window is successively shifted over the entire map of elevation to produce a continuous slope or aspect map. Insert (a) of figure 7 shows the derived map of aspect for the area. Note that a 'slope map' of any surface represents the first derivative of that surface. For an elevation surface, slope depicts the rate of change in elevation. For an accumulation cost surface, its slope map represents the rate of change in cost (i.e. a map of marginal cost). For a travel-time overlay, its slope map indicates the relative change in speed and its aspect map identifies the direction of travel at each location. The slope map of an existing topographic slope map (i.e. second derivative) will also characterize surface roughness.

The creation of a 'profile map' uses a window defined as the three adjoining points along a straight line oriented in a particular direction. Each set of three values can be regarded as defining a cross-sectional profile of a small portion of a surface. Each line of data is successfully evaluated for the set of windows along that line. This procedure may be conceptualized as slicing a loaf of bread, then removing each slice and characterizing its profile (as viewed from the side) in small segments along its upper edge. The centre point of each three-member neighbourhood is assigned a value indicating the form of the profile at that location. The value assigned can identify a fundamental class of profile (e.g. an inverted 'V' shape indicates a ridge or peak) or it can identify the magnitude, in degrees, of the 'skyward angle' formed by the intersection of the two line segments of the profile. The result of this operation is a continuous map of a surface's profile as viewed from a specified direction. Depending on the resolution of an elevation map, its profile map could be used to identify gullies or valleys running east/west (i.e. 'V' shape as viewed from the east or west profile) or depressions (i.e. 'V' shape in orthogonal profiles).

The other group of neighbourhood operations is that which summarizes thematic values. Among the simplest are those involving the calculation of summary statistics associated with the map categories occurring within each neighbourhood. These statistics might include, for example, the maximum income level, the minimum land value or the diversity of vegetation types within a quarter of a mile radius (or, perhaps, a 5-min walk) of each target point. Insert (b) of figure 7 shows the diversity of cover types

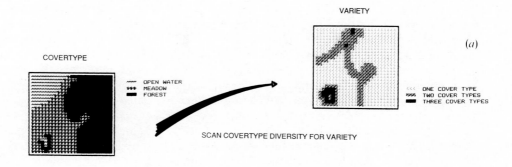

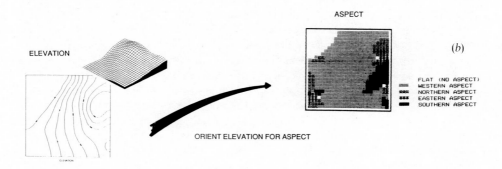

Figure 7. Characterizing neighbourhoods. These operations summarize the attributes occurring in the vicinity of each location. Insert (a) is a map of topographic aspect generated by successively fitting a plane to neighbourhoods of adjoining elevation values. Insert (b) is a map of cover type diversity generated by computing the number of different cover types in the immediate vicinity of each map location.

occurring within the immediate vicinity of each location. Other thematic summaries might include the total, the average or the median value occurring within each neighbourhood, the standard deviation or variance of those values, or the difference between the value occurring at a target point and the average of those points surrounding it.

None of the characteristics of neighbourhoods described so far relates to the area occupied by the map categories within each neighbourhood. Similar techniques might be applied, however, to characterize neighbourhood values which are weighted according to areal extent. One might compute, for example, total land value within 3 miles of each target point on a per-acre basis. This consideration of the size of the components also gives rise to several additional neighbourhood statistics including mode (the value associated with the greatest proportion of neighbourhood areas), minority value (the value associated with the smallest proportion of a neighbourhood area) and uniqueness (the proportion of the neighbourhood area associated with the value occurring at the target point itself).

Another locational attribute which might be used to modify thematic summaries is the cartographic distance from the target point. While distance has already been described as the basis for defining a neighbourhood's absolute limits, it might also be used to define the relative weights of values within the neighbourhood. Noise level, for example, might be measured according to the inverse square of the distance from surrounding sources. The azimuthal relationship between neighbourhood location and the target point may also be used to weight the value associated with that location. In conjunction with weighting by distance, this gives rise to a variety of techniques of spatial sampling and interpolation. For example, 'weighted nearest-neighbour' interpolation of data for lake bottom temperatures assigns a value to an unsampled location as the distance-weighted average temperature of a set of sampled points within its vicinity.

7. Cartographic modelling

The preceding discussion has developed a typology of fundamental procedures for map processing and described a set of reclassifying, overlaying, distance and neighbourhood operations common to a broad range of higher techniques of map analysis. By systematically organizing these primitive operations, the basis for a generalized cartographic modelling approach can be identified. This approach accommodates a variety of analytic procedures in a common, flexible and intuitive manner analogous to the mathematical structure of conventional algebra.

As an example of some of the ways in which fundamental operations for map processing might be combined to perform more complex analyses, consider the cartographic model outlined in figure 8. The format uses boxes to represent encoded and derived maps and lines to represent primitive map processing operations. The structure of the flowchart indicates the logical sequencing of operations on the mapped data that progresses to the desired final map. The simplified cartographic model shown depicts the siting of the optimal corridor for a highway with reference to only two criteria: an engineering concern to avoid steep slopes and a social concern to avoid visual exposure. Implementation of the model using the pMAP system requires less than 25 lines of code. The execution of the entire model using the data base in the previous figures requires less than 6 min on an IBM PC/AT microcomputer.

Given a map of topographic elevation and one of land use, the model allocates a minimum-cost alignment of a highway between two predetermined termini. Cost is measured not in dollars but in terms of locational criteria. The right-hand portion of the flowchart develops a 'discrete cost surface' in which each location is assigned a relative cost based on the particular combination of steepness and exposure occurring at that location. For example, those areas that are flat and not visible from houses would be assigned low values, whereas areas that are on steep slopes and visually exposed would be assigned high values. The discrete cost surface is used as a map of relative barriers for establishing an 'accumulated cost surface' from one of the termini to all other locations within the mapped area. The final step locates the other terminus on the accumulated cost surface and identifies the minimum cost route as the steepest downhill path along the surface from that point to the bottom (i.e. the other end point).

In addition to the benefits of efficient data management and of automating cartographic procedures, the modelling approach to computer-assisted map analysis has several other advantages. Foremost among these is the capability of dynamic simulation' (i.e. spatial 'what if' analysis). Thus the model for siting highways could be executed for several different relative weightings of the engineering and social criteria,

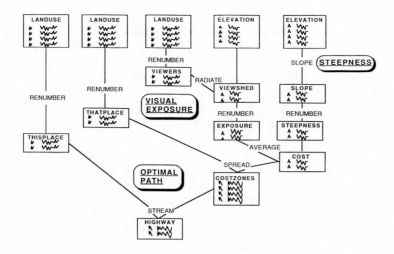

Figure 8. Cartographic modelling. This simplified cartographic model depicts the siting of the optimal corridor for a highway with reference to only two criteria: an engineering concern to avoid steep slopes and a social concern to avoid visual exposure. In a manner similar to conventional algebra, this process uses a series of map operations (indicated by lines) to derive intermediate maps (indicated by boxes), leading to a final map of the optimal corridor.

for example, if steepness of terrain is more important or if visual exposure is twice as important. It could also identify where the minimum cost route changes or, just as importantly, where it does not change. From this perspective, the model 'replies' to inquiries from users, rather than 'answering' them: providing information, rather than tacit decisions.

Another advantage of cartographic modelling is its flexibility. New considerations may be easily added and existing ones refined. For example, the non-avoidance of bodies of open water in the highway model is a major engineering oversight. In its current form, the model favours construction on lakes, as they are flat and frequently not visually exposed. This new requirement can be readily incorporated by identifying open water bodies as absolute barriers (i.e. infinite cost) when constructing the accumulation cost surface. The result will be routing of the minimal cost path around these areas of extreme cost.

Finally, cartographic modelling provides an effective structure for communicating both considerations of specific applications and fundamental procedures for processing. The flowchart provides a succinct format for communicating the logic, assumptions and relationships embodied in an analysis. Such a presentation encourages the active involvement of decision makers in the analytic process.

8. Conclusion

The preceding discussion has developed a typology for computer-assisted map analysis and described a set of independent analytical operations common to a broad range of applications. By systematically organizing these operations, the basis for a generalized cartographic modelling approach is identified. This approach accommodates a variety of analytic procedures in a common, flexible and intuitive manner, analogous to the mathematical structure of conventional algebra.

Management of land has always required spatial information as its cornerstone. However, procedures for integrating this information fully into decision making have been limited. Traditional statistical approaches seek to constrain the spatial variability within the data. Sampling designs involving geographical stratification attempt to reduce the complexity of spatial data. However, it is the spatially-induced variation of mapped data and their interactions that most often concern land managers. This approach retains the quantitative aspects of the data necessary for most decision-making models, but lacks spatial continuity. On the other hand, the approach through drafting is spatially precise, but limited by both its non-quantitative nature and its laborious procedures. In most instances, a final drafted map represents an implicit decision involving only a few policy strategies, rather than a comprehensive presentation of information.

Computer-assisted map analysis, on the other hand, involves quantitative expression of spatially-consistent data. In one sense, this technology is similar to conventional map processing involving traditional maps and drafting aids such as pens, rub-on shading, rulers, planimeters, dot grids and acetate sheets for light-table overlays. In another sense, these systems provide advanced analytic capabilities, enabling managers to address complex issues in entirely new ways. Within this analytic context, mapped data truly becomes spatial information for effective decision making.

References

Berry, J. K., 1986 a, A mathematical structure for analyzing maps. *Environmental Management* (in the press).

Berry, J. K., 1986 b, Geographic information systems (Part III): learning computer-assisted map analysis. *Journal of Forestry*, **84,** 39–43.

Berry, J. K., 1986 c, Geographic information analysis workshop workbook. *The Academic Map Analysis Package (aMAP) Instructional Materials*, 9th edition (New Haven, Connecticut: Yale University, School of Forestry and Environmental Studies).

Berry, J. K., and Tomlin, C. D., 1982 a, Cartographic modelling: computer-assisted analysis of effective distance. *Proceedings of the 8th Symposium on Machine Processing of Remotely Sensed Data*, Purdue University, West Lafayette, Indiana, pp. 503–510.

Berry, J. K., and Tomlin, C. D., 1982 b, Cartographic modelling: computer-assisted analysis of spatially defined neighborhoods. *Proceedings of Energy Resource Management Conference* (Chicago: American Planning Association), pp. 307–319.

SIS, 1986, The Professional Map Analysis Package (pMAP) user's manual and reference. Spatial Information Systems, Incorporated, Omaha, Nebraska.

Tomlin, C. D., 1983 a, Digital cartographic modelling techniques in environmental management. Doctoral dissertation, Yale University, School of Forestry and Environmental Studies, New Haven, Connecticut.

Tomlin, C. D., 1983 b, A map algebra. Harvard Computer Graphics Conference, Harvard University, Cambridge, Massachusetts (unpublished handout at conference).

Tomlin, C. D., and Berry, J. K., 1979, A mathematical structure for cartographic modelling in environmental analysis. *Proceedings of the 39th Symposium*, American Congress on Surveying and Mapping, Falls Church, Virginia, pp. 269–283.

Computer-assisted relief representation

Pavao Stefanovič[*] and Koert Sijmons[*]

ABSTRACT

The different possibilities of relief representation by means of digital elevation models (DEMs) are discussed. Emphasis is placed on the manipulation of data to obtain different types of relief representations (oblique hill shading, slope representation, contour lines and layer tints), demonstrated with a map fragment. The suitability of DEMs depends to a high degree on the reliability of the selected sampling method.

Because relief is a three-dimensional phenomenon, its representation in two-dimensional maps poses some difficulties. The fact that the whole range of graphic means–points (spot heights), lines (contours), areas (hill shading)–is traditionally used to depict relief illustrates the problem. Moreover, qualitative as well as quantitative methods of relief representation are used, for which numerous cartographic techniques and various types of equipment are utilized.

Contour lines are used most frequently to represent relief, but other means, such as oblique and vertical hill shading, slope mapping, layer tinting, oblique views, etc, have gained in popularity. Their construction is traditionally based on contour lines which are originally obtained by field survey, manual interpolation within a grid of points, or direct photogrammetric plotting. These are typically time-consuming and labour intensive activities requiring highly skilled and well trained personnel.

Digital techniques and computer assistance have opened new possibilities for overcoming these difficulties, especially when a variety of relief representations are intended for a certain region. In such cases, contour lines are no longer the basic data source because their role is taken over by digital elevation models (DEMs).

DATA SOURCES FOR DIGITAL ELEVATION MODELS

DEMs can be defined as digital representations of the terrain in a form suitable for computer processing. In addition to the input data (X, Y and Z coordinates given directly or indirectly) for a set of selected surface points, processing requires appropriate computer programs.

The main bottleneck in using DEMs is data acquisition. It is traditionally a time-consuming and costly operation and DEM technology has brought little change in this respect. Once the data are available in appropriate form, however, any type of relief representation can be obtained more economically and more rapidly than ever before and–in addition–relief analysis can be carried out directly and efficiently.

The following are important for DEM data acquisition: data source, equipment, sampling pattern (points), point density, and the data fidelity required. There are three main data sources: field surveys, photo or other images, and existing maps. The data sources directly govern the selection of equipment. When sampling is done in the field, survey instruments must be used. If the data source is stereo imagery (terrestrial, aerial or satellite), photogrammetric equipment is needed, while existing maps require the use of cartographic digitizing equipment. The most popular sampling patterns are contours, profiles, grids, and breaklines and points, but a combination of these patterns is of course possible in any one sampling technique.

Field surveys give the greatest accuracy for sample points. The high cost per point, however, necessitates a low point density–which adversely influences the accuracy of interpolated points. With low point density, the only possible efficient sampling pattern uses points indicating significant changes in relief. The selection of such points obviously requires highly trained personnel.

Sampling based on stereo images and photogrammetric equipment allows a much higher density of points and variety of patterns. Sampling accuracy is restricted, however, by the scale and quality of the stereo images. Contours, profiles and grids–with some lines supplemented by break lines and points–are all used. The preferred patterns are either the traditional ones (such as contours) or those which require minimum terrain interpretation (such as grids and profiles).

In addition to the manual sampling mode, photogrammetric data acquisition also permits both semi-automatic and automatic sampling. Contour lines are suitable for manual methods, while grids and profiles are appropriate for semi-automatic methods. Planimetric positioning is automatically controlled, while height settings are completed manually.

Instruments with automatic components (stereo correlators) permit completely automatic data sampling. The cost per point drops so drastically that a

---

* Cartography Department, ITC

very high point density can be used. On the other hand, the accuracy of individual points is lower than can be obtained with manual or semi-automatic methods.

For a large part of the earth's surface, there are existing maps which show contours and spot heights. They are a valuable and relatively inexpensive source of data. The fidelity of the DEM, however, is extremely sensitive to the character of the data source. The map scale, the quality of the original survey and cartographic presentation do not allow much variety in data sampling and influence the DEM's quality. The use of relatively inexpensive cartographic manual digitizers involves manually following contour lines, which is–of course–a very tedious and time- consuming procedure. The recent development of raster scanners–and their impact on contour digitizing–opens new possibilities for speeding up this process. Unfortunately, raster scanners require substantial investment.

Data fidelity is the most important single factor in DEM data acquisition. This term applies not only to the accuracy of the originally sampled points but also to any point to be derived from the original data set. It is obviously influenced by equipment, data storage methods, sampling pattern, measuring accuracy, and the density and method of interpolation. The fidelity of relief representation is traditionally judged not only by criteria of accuracy, however, but also by the manner in which it reflects the character of the relief. This second aspect, together with aesthetic requirements of relief representation, makes the control of fidelity especially difficult. In small-scale representations, even accuracy is sometimes sacrificed to obtain what is considered a faithful representation of relief. Considering the complexity of these problems, there will be scarcely any DEM data perfectly suitable for any desired type of representation. The obvious solution, then, is post- processing of data to prepare them for specific applications.

ARCHIVING DEM DATA

Before the data are permanently stored, they must be cleaned of gross errors and unwanted noise. The large volume and the digital character of the data make this operation quite difficult. The detection of gross errors requires some effort; automatic analysis of gradients and inspection of the derived contours or oblique views seem to give the most promising results. The automatic data acquisition methods require special post-processing to eliminate the intolerable noise. This usually involves some sort of smoothing.

Clean data are ready for archiving, but two major problems arise: how to store the very large amount of data, and how to structure the data storage to make it easily accessible and useable for various purposes.

There are data compression techniques which permit efficient data storage. Compression (and later decompression) routines increase computer processing time, but recent spectacular increases in the capacity of computer storage media, and decreases in their prices, have made data compression a less important topic.

The importance of the structure of stored data has not diminished, however, because the structure influences all further processing. The regular grid is considered as the most generally applicable data structure and almost all existing archiving systems use it—thus achieving a certain standardization. This, of course, requires data conversion in all cases where the sampling pattern deviates from a grid.

Despite the apparent standardized input for processing DEM data into any desired output form, the fidelity characteristics of archived data are strongly influenced by the data sampling methods. Because they differ so much, the quality of computer-assisted relief representations may be disappointing; many results confirm this, especially if the sampling peculiarities are not resolved. This is not an easy task, however, because the quality of archived data is usually assessed only with respect to accuracy and not to fidelity. The assessment of fidelity is very difficult in any event because it depends on the scale of representation, type of representation and the attitude of the map user–which all might be unknown at the moment of sampling.

Standardization of data processing has enormous practical advantages which should not be neglected, but it creates a dilemma for the user. He must either adjust his standards to the actual computer output or upgrade the output in the post-processing stage.

The first solution is characterized by low cost. It should be kept in mind that the accuracy of such a presentation can be higher than in any comparable manually-prepared map. The product can be upgraded with an interactive graphic system. This procedure is very effective, but time- consuming and costly. It can be easily adapted to any requirement because the operator decides in every situation what to do and in which way. Smoothing techniques can sometimes be applied. Although it is typically batch processing, the results of smoothing must be visually inspected each time because the effect is not easily predictable.

REPRESENTATION TECHNIQUES

The availability of drafting machines is an essential prerequisite for the practical application of computer-assisted relief representation. Line plotters have been used for this purpose for a long time, but they have only recently reached the drafting speeds that make them commercially attractive. The line quality in the drafting products has reached such a high standard that they satisfy the most demanding requirements of the reproduction industry.

The real breakthrough, however, has been

achieved by the latest developments in raster plotters. They not only offer much higher drafting speeds, but also permit types of relief representation which are either impossible or very difficult with line plotters.

Graphic screens have also had an important impact. The raster refresh colour screens allow any type of instant relief representation without any wastage of materials. Their drawbacks are high cost, small size and insufficient resolution, but these are often compensated by completely new and exciting possibilities such as continuously changing oblique views of the terrain.

Some of the most popular types of computer-assisted relief representation are described below. An existing map at scale 1:100 000 was the data source, and digitizing was manual. The digitized contours are shown in Figure 1.

Conversion into a regular grid was done with the help of the Hifi program developed at the Munich Technical University. The combination of the sampling technique and interpolation method (finite element) is most probably not the optimal one. The derived regular grid with a density of approximately 1 mm in map scale was then used to produce all relief representations that follow.

OBLIQUE VIEWS

Although oblique views were the earliest type of relief representation, they were almost abandoned later except as artists' impressions. The expenditures of time and money for manual preparation of accurate oblique views are prohibitively high.

The use of computers has changed this situation drastically. Preparation of such views, especially using a regular grid as input, is rather quick and straightforward. Numerous programs have been developed for this purpose using various types of projections (perspective, axonometric, etc). Inherent problems such as the "hidden line" have been successfully solved.

The so-called "fishnet representation" is especially popular. The surface points are interconnected by straight lines to make the surface visible (see Figure 2). The critical aspect in producing these is the point of view. It is not always clear in advance (and especially if experience is lacking) which point will provide the most suitable view. It is often decided by trial and error. Drafting such views using line plotters is rather time-consuming because of the large number of lines. Inexpensive electrostatic raster plotters can be used if high line quality is not required. A more sophisticated version of an oblique view allows various other features to be superimposed onto the relief surface, which–if properly done–gives a very realistic view of the terrain. Three-dimensional representations of such features must then be available in the data set, however. It is relatively easier to work with digital remote sensing data, which results in a variegated oblique view of the terrain.

HILL SHADING

Hill shading is traditionally a more artistic than objective or accurate relief representation. Because it is considered indispensable in some types of maps, computer-assisted hill shading must be included in any comprehensive production system.

Analytical hill shading is the basis of computer-assisted methods. It was originally developed as a more objective method, but its practical application was hardly feasible in a manual production environment because of the large number of computations. Advances in computers revived interest in it, but did not make it more attractive in practice. Only since the recent development of high resolution photo-optical raster plotters and high capacity storage media has there been a real basis for the inclusion of such methods in production.

The principles of analytical hill shading are rather simple. First, a surface element has to be formed. In this case, a rectangular grid represents the input data. Such an element is logically determined by four neighbouring grid points (see Figure 3). After defining the position of the observer (0) and the source of illumination (S) we can define three angles:

E - the angle between the normal to the surface and the source of illumination

A - the angle between the direction to the observer and the source of illumination

B - the angle between the normal to the surface and the direction to the observer

Computation of the angles is relatively simple after the positions of the source of illumination and the observer have been fixed. To obtain the shaded relief, a grey level code (g) must be assigned to each surface element as a function of the angles $g = f(E, A, B)$. The function can assume diverse forms and the number of possibilities is large. The complex forms of the function (sometimes based on reflectance theory), however, usually do not give results better than simple ones.

The size of the surface element must be selected to match the resolution of the output raster plotter. The area of the surface element should not exceed 0.1 × 0.1 mm in map scale; otherwise the pixel pattern becomes visible. This means that, except with high resolution plotters (photo-optical plotters), the resulting output has to be photographically reduced to obtain a reasonable graphic quality. Moreover, such a small size of the element usually requires densification of the input DEM, unless it is already very dense.

Figure 4 shows computer-assisted hill shading of the test area. The input grid mesh was 1 mm in map scale. The result was printed with the Optronics electro-optical raster plotter in continuous tone and with a resolution of 0.1 mm. The film thus obtained had to be mechanically screened for printing. Film development and screening are difficult to control accurately and the resulting grey tones may not correspond exactly to the original design. A much better procedure is

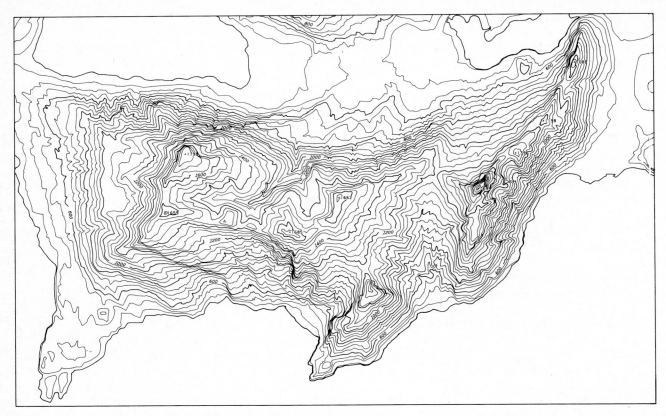

FIGURE 1 Contour lines from 1:100 000 topographic map of Switzerland (courtesy of Swiss Topographic Service)

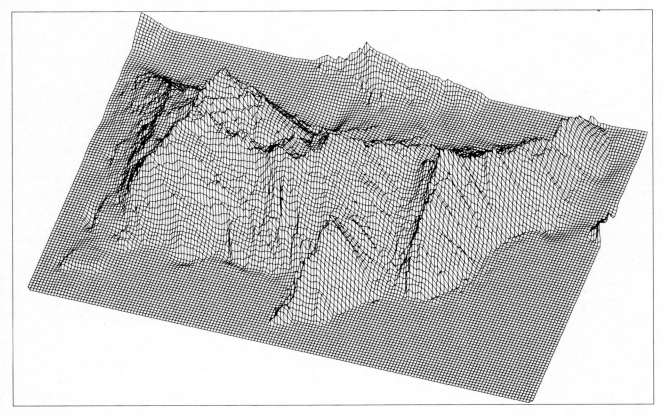

FIGURE 2 Isometric view

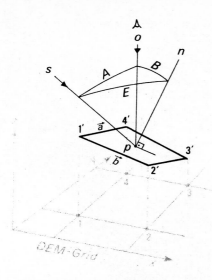

FIGURE 3 Geometric model for illumination of surface element

to produce digital screening directly with a raster plotter, which gives exact control of grey tones. Much finer resolution has to be used, however, probably at least 0.025 mm. The grey level code was computed for the Optronics plotter in the following manner:

$$g = R \cos E$$

where

$$R = 0 \text{ if } E \leqslant 0$$
$$R = 255 \times f \text{ if } E \geqslant 0$$

For the particular raster plotter code, 255 represents white and 0 represents black. Factor f is an exponential function of E which has to be adapted to the characteristics of the film and developing procedure.

SLOPE REPRESENTATION

Manual slope mapping is, without doubt, a very tedious and inaccurate procedure. Here the operational speed, accuracy and ease of computer-assisted methods are so superior that hardly anyone questions their use.

The principles of computer-assisted slope representation are more simple than hill shading. Using Figure 3 again, it is obvious that slope is directly determined by angle B if the observer is at the zenith. Although continuous slope representation is easy to obtain by assigning the grey level code proportional to the angle B (where g = RB), it is usual practice to represent the slopes grouped into classes. The class rank number can then be associated with the appropriate grey level or colour code through a Look Up Table and printed directly on a raster plotter.

An example of a slope map of the test area is shown in Figure 5. The input grid characteristics are the same as in hill shading; densification was therefore also necessary. Grey level codes were assigned to

selected slope classes through a Look Up Table. The raster plotter printing, however, was done directly on digital screens, which required a plotter resolution of 0.050 mm but delivered printable originals.

CONTOUR LINES

For all of the preceding relief representations, there is little doubt that computer-assisted methods are far superior to manual methods. For contour lines, however, this is not so. When contour lines are the only relief representation and maps with diverse contour intervals are not planned, it is especially difficult to justify the computer-assisted approach.

Although the contour lines derived from the DEM are, as a rule, sufficiently accurate, the peculiarities of any specific interpolation method produce artifacts which are not always tolerable. The remedy is either inclusion of morphologic lines or interactive post-processing–both requiring additional human effort, depending on the complexity of the terrain.

There are two approaches to contour line interpolation. The first one, known as contour threading, has been used for a number of years and delivers results in vector format. The principle is illustrated in Figures 6A and 6B. Each contour line is followed individually through the grid. Finally, all intersection points with grid lines are stored in proper sequence. The grid mesh must be sufficiently fine so that, despite straight line segments between the individual stored points, the whole contour has a smooth appearance. This usually requires densification of the input DEM.

The result can be output to a line plotter. The programming for this type of contour production is not very simple; difficulties include specific height relationships among mesh corner points and determining if all possible contours have actually been found. These difficulties can cause a fatal failure.

The second approach uses raster format and a raster plotter. The whole DEM has to be densified down to the raster plotter resolution (pixel size). The interpolation can then be completed, scan line-by-scan line, in a single pass. The programming is also extremely simple and requires only testing of elevations of the corner points of each pixel. If at least one corner point elevation is smaller or equal to the contour elevation, and at the same time at least one is larger, the pixel lies on the contour. This principle is illustrated in Figures 6A and 6C. The procedure is quick, failure is not possible, and the result can be plotted directly on a raster plotter.

The results of such contouring are shown in Figure 7, using the same DEM input as in the previous examples.

LAYER TINTS

Efficient computer-assisted layer tinting is very similar to contour line interpolation in raster format (see Figures 6A and 6D). The only difference is that

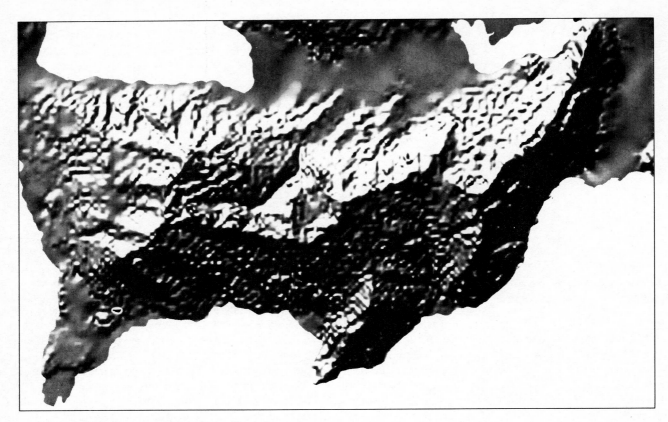

FIGURE 4 Oblique analytical hill shading

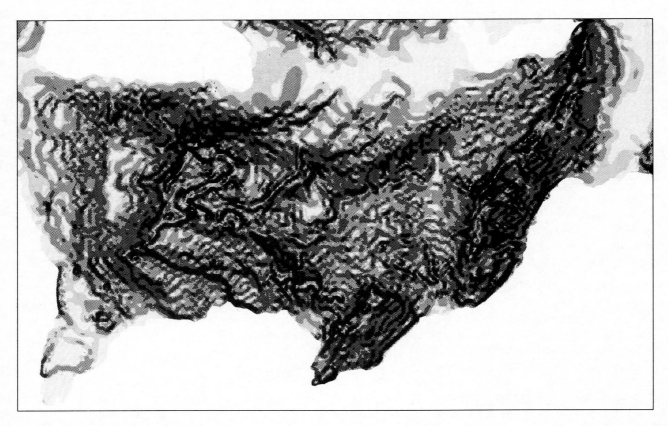

FIGURE 5 Analytical slope map

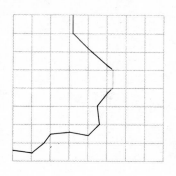

FIGURE 6A DEM patch used as example in Figures 6B, 6C and 6D

FIGURE 6B Interpolated vector contour line

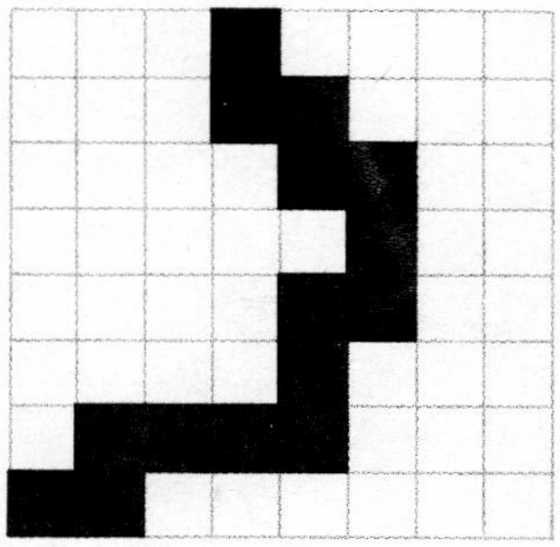

FIGURE 6C Interpolated raster contour line

FIGURE 6D Layer tints

detection of each contour pixel acts as a switch to a lower or higher elevation layer. In the output, the appropriate layer code is assigned to all pixels between two neighbouring contours. Through a convenient Look Up Table, such layer codes can be converted into corresponding grey level or colour codes and printed on a raster plotter.

Figure 8 shows the layer tints of the test area. Eight layers were produced by digital screening.

The line plotter and cut-and-peel technique can also be used for layer tinting, in a manner very similar to manual methods. This obviously requires a much higer percentage of manual work than the method described above.

ACKNOWLEDGEMENTS

The authors gratefully acknowledge the contribution of Mr J P R van den Worm in digitizing the original contour lines and Mr E Shamai in preparing the map fragments.

BIBLIOGRAPHY

American Society of Photogrammetry. 1978. Proc of the Digital Terrain Models (DTM) Symposium, St Louis, Missouri, May 9-11.

Batson, R M, E Edwards and E M Eliason. 1975. Computer generated relief images. Journal of Research, U S Geological Survey, Vol 3, No 4, July- August, pp 401-408.

Brassel, K. 1973. Modelle und Versuche zum automatischen Schraglichtsschattierung. Ph D dissertation Geography Dept, University of Zurich, Klosters, Switzerland.

Douglas, D. 1971. VIEWBLOCK: A computer program for constructing perspective view block diagrams. Revué de Geographie de Montreal, Vol 26, pp 102-104.

Horn, B K P. 1982. Hill shading and the reflectance map. Geo-Processing, 2: pp 65-146.

Imhof, E. 1965. Kartografische Gelandedarstellung. Berlin, W de Gruyter and Co.

Peucker, T K and D Cochrane, 1974. Die Automation der Reliefdarstellung - Theorie und Praxis. International Yearbook of Cartography, Vol 14, pp 128- 139.

Robinson, A H. 1961. The cartographic representation of the statistical surface. International Yearbook of Cartography, Vol 1, pp 53-63.

Silar, F. 1972. Das digitale Gelandemodell - Theorie und Praxis. Vermessungstechnik, Vol 20, No 9, pp 327-329.

Sprunt, B F, 1969. Computer-generated halftone images from Digital Terrain Models. MSc Thesis, Department of Mathematics, Univ Southampton.

Wiechel, H. 1978. Theorie und Darstellung der Beleuchtung von nicht gesetzmassig gebildeten Flachen mit rucksicht auf die Bergziehung. Civielingenieur, Vol 24, pp 335-364.

Yoeli, P. 1965. Analytical hill shading. Surveying and Mapping, Vol 25, No 4, Dec, pp 573-579.

Yoeli, P. 1971. An experimental electronic system for converting contours into hill-shaded relief. International Yearbook of Cartography, Vol 11, pp 111-114.

RESUME

Les différentes possibilités de la représentation du relief au moyen des modèles numériques du terrain (MNT) sont discutées. L'accent est mis sur la manipulation des données pour obtenir différents types de représentation du relief, (projection d'ombres obliques, représentation de pentes, courbes de niveau et couche de teintes) et démontré avec un fragment de carte. L'emploi approprié du MNT dépend essentiellement de la fiabilité de la méthode d'échantillonnage.

RESUMEN

Se discuten las diferentes posibilidades de representacion de relieves a traves de modelos de elevacion digitales (DEMs). El enfasis esta en la manipulacion de informacion para la obtencion de diferentes tipos de representacion de relieves (sombreo oblicuo de montañas, representacion de pendientes, cotas y color de estratos) mostrados en un fragmento de mapa. La idoneidad de DEMs depende en un alto grado de la exactitud de los metodos de muestreo seleccionados.

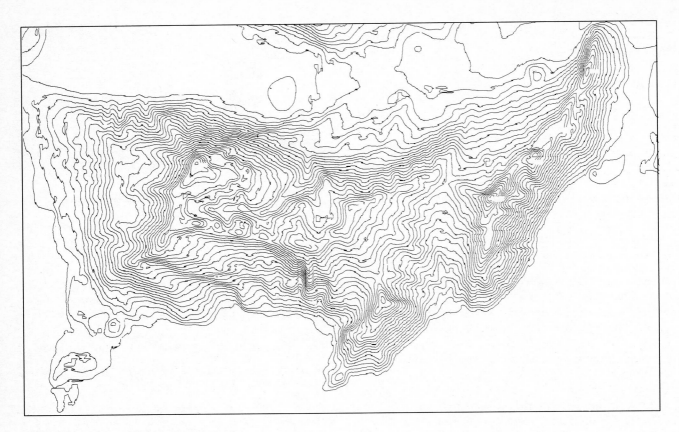

FIGURE 7 Raster generated contour lines

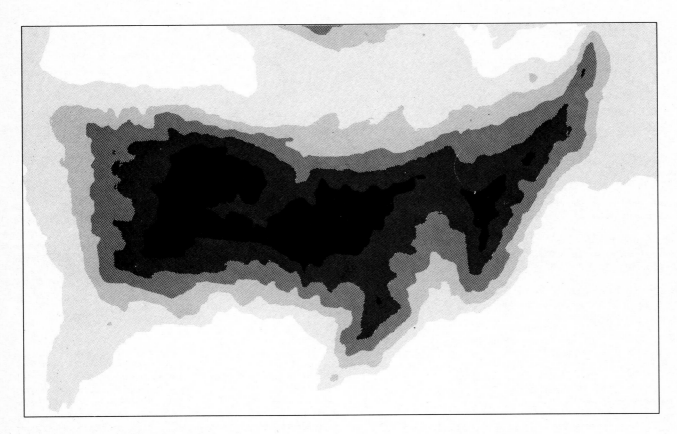

FIGURE 8 Raster generated layer tints

SECTION 4
Data Quality

Overview

This section addresses the problem of assessing errors in GIS. The first of the three papers, by Drummond, provides a mathematical framework for attaching a data quality parameter to GIS output. Wehde illustrates the relationship between grid cell size and map error in a raster-structured GIS. The last article, written by Newcomer and Szajgin, establishes a method for calculating the accuracy of composite overlay maps in a GIS.

Suggested Additional Reading

Bailey, R.G. 1988. Problems With Using Overlay Mapping for Planning and Their Implications for Geographic Information Systems. *Environmental Management*. 12(1):11–17.

Chrisman, N.R. 1984. The role of Quality Information in the Long Term Functioning of a Geographic Information System. *Cartographica*. 21(2/3):131–139.

Walsh, S.J., D.R. Lightfoot, and D.R. Butler 1987. Recognition and Assessment of Error in Geographic Information Systems. *Photogrammetric Engineering and Remote Sensing*. 53(10):1423–1431.

A framework for handling error in geographic data manipulation

Jane Drummond[*]

ABSTRACT

This article addresses the problem of attempting to provide an attached quality parameter to information provided by a geo information system (GIS). Error propagation theory can be applied to the mathematical model approach but not to the logic model approach. A procedure is proposed for generating quality parameters for logic models. Methods of deriving quality parameters for point sample data and graphic boundary map data are described. The importance is emphasized of being able to estimate numerically the accuracy of the data used as input to a GIS.

In a geo information system (GIS), it can be expected that data will come from several sources, such as field observations (measurements), maps, and satellite or airborne sensors. These data are processed to provide further geo information for financiers, planners, politicians, agriculturalists, etc. Currently, it is rare for the information provided by a GIS to have an attached quality indicator. If the information is going to be the basis of development investment, however, users have to be aware of its reliability and the resulting consequences. The data stored in a geo information system include rainfall, soil depth, elevation, vegetation, temperature, and similar information. These are all functions of observations and represent random variables. The results of any data processing within a geo information system are therefore also random variables and we may thus use statistical inference to derive quality indicators.

One possible quality indicator is percentage probability, ie, probability expressed in percentage terms. (Probability is often expressed as value ranging from 0 to 1, and must be converted into this form for computations.)

It would be useful to all concerned if we were capable of making such statements as: "there is a 92 percent probability that this land parcel is very suitable for growing irrigated rice", or "there is a 56 percent probability that that land parcel is very suitable for growing irrigated rice".

Clearly, limited resources might more sensibly be invested in the first land parcel than the second. It is the purpose of this article to outline some possibilities for achieving this capability. The statistical techniques introduced are in no way, novel, but for practicing cartographers who may be the GIS managers of the future, it is hoped that this article will make some contribution to the ongoing discussions on quality in the digital environment.

MATHEMATICAL AND LOGIC MODELS

Two examples of geo information supplied by a GIS are annual soil loss per hectare and crop suitability rating. Although both may be used for assessing a land parcel, they are the result of thinking about and processing geographic data following two different approaches.

The first approach (that for calculating annual soil loss/hectare) uses a mathematical (or arithmetic or functional) model, such as the "universal soil loss equation" [4]. This approach provides a numeric "answer" which may aid in decision making. The second approach uses a logic (or heuristic) model, such as those developed for the Agusan River Basin of the Philippines [10], and can be used to determine crop suitability ratings for several tropical crops. Again, help in decision making may be achieved.

The models used by land surveyors and photogrammetrists to determine coordinates, and subsequently other measurements, are mathematical models. In mathematical models, unknown variables (*eg*, phi rotation or annual soil loss) are expressed as a function of observations estimated at an earlier stage. Conventional error propagation theory [14] is available to determine the quality of results derived from the application of such models. It is a prerequisite, however, that only random errors influence the result. Some mathematical models used in the geo sciences may contain systematic error of an unknown magnitude. Furthermore, the models may not result from a rigorous application of dimensional analysis (for example) or other techniques used in the physical sciences. There may be very little known about the problem at hand, but experience may have spawned a mathematical model which is valuable in practice. These models may be called "empirical mathematical models". The universal soil loss equation is one such empirical mathematical model.

Logic models are not based on mathematical formulae and do not lend themselves to error propagation theory, but they do appear to be popular in GISs [8, 13], because in many application fields they more closely represent the workings of an "expert" than do mathematical models. A logic model can be used, for example, to estimate the suitability rating of a land parcel for growing irrigated rice. The suitability rating may be defined in terms of "very suitable", "suitable", or "unsuitable" for irrigated rice. The rating will have been derived from logical rules such as a land parcel being very suitable only if the rainfall is

* Department of cartography, ITC

greater than 700 mm per year, if the elevation is less than 200 m, and if the soil depth is greater than 100 cm. There are no recognized techniques for quality estimation when a logic model is used.

STATISTICAL ASSUMPTIONS

Mathematical models encapsulate rules that can be used to make estimates of unknown variables from existing observations or other variables. The best estimate, or most probable value, of a variable should derive from the weighted mean of several observations, but in practice entirely acceptable alternatives may be used as estimates and the most obvious of these is the mean. Because observations vary when they are repeated, the best estimate, or most probable value, has a probability distribution associated with it whose shape, or spread, gives an indication of the quality of the estimate. Standard deviation (SD or σ) gives a statistical indication of the spread of this distribution and can therefore be used as a quality indicator. It is thus possible, according to classical statistics, to link SD with the concept of the probability that a variable's value is within a certain interval around the estimate. Users of a GIS may feel more comfortable with a quality indicator based on probability than one based on SD.

To exploit the linkage indicated in the preceding paragraph, some assumptions must be made; these are: (1) that the estimate is unbiased; (2) that observations are uncorrelated; (3) that the probability distribution of observations is known; (4) that the error related to an observation is normally distributed; and (5) that the variance must be finite.

SD assumes a central role in quality considerations and can be determined from established procedures operating on the dispersion of observations around an estimate or, alternatively, from experience, educated guesses, or other considerations. A link between SD and probability must therefore be established. The techniques for deriving an estimate by confidence intervals will be used.

An estimate ($\bar{x}$) of the unknown value (θ) is derived from several observations ($x_1, x_2, x_3 \ldots x_i$). A confidence interval for $\bar{x}$ can be computed from the observations such that θ is known to fall somewhere between the lower (L) and upper (U) bounds of the confidence interval with a preselected (and usually high) probability, *eg*, 95 percent (0.95) or 99.7 percent (0.997). This relationship can be stated as the inequalities:

$$p\,[L < \theta < U] = 0.95$$

or

$$p\,[L < \theta < U] = 0.997$$

If the estimate of a value is set to zero, then the bell-shaped curve of the normal (Gaussian) probability distribution of the observations used to determine that estimate stretches from $-\infty$ to $+\infty$ (Figure 1) and the area under it is a function of all possible observations contributing to the estimate. A property of such a curve is that if two x-values are selected and they (for example) generate a sub-area equal to 90 percent of the whole area under the curve, then the sub-area encloses 90 percent of all possible observations. It follows that an excluded sub-area of 10 percent exists; this is termed (a). The included area, in percentage terms, is thus 100(1-a) percent.

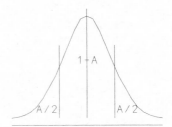

FIGURE 1 Normal probability distribution showing a-values

If the two values selected on the x-axis are symmetrical about the central value, then the excluded sub-area (a) is in two symmetric parts, each of area (a)/2, found in the tails of the curve. The upper x-value (sometimes called the upper F-value) may be depicted by $z_{a/2}$ and the lower intersection point (lower F-value) by $-z_{a/2}$. In the case of intersection points being symmetrically positioned about the central value, the absolute values of $z_{a/2}$ and $-z_{a/2}$ are the same. It can thus be understood that associated with each $z_{a/2}$ or $-z_{a/2}$ there is an excluded percentage of the total area under the curve of 100(a/2) percent. Also associated with each $z_{a/2}$ or $-z_{a/2}$ is an included percentage of the total area under the curve of 100((1-a)/2) percent. Normal probability tables (*eg*, [9], appendix B) may be used to determine the areas under different parts of the curve associated with different F-values scaled from 0 (0 percent) to 1 (100 percent). A few such values are shown in Table 1.

The F-values are set to be multipliers of the standard deviation of the variable under consideration; thus if the F-value is 1, there is then a 68 percent probability that the value θ will fall between $\bar{x} - 1 \times$ SD and $\bar{x} + 1 \times$ SD (see Table 1). This can be restated as:

$$P\,[\bar{x} - z_{a/2} \times \sigma < \theta < \bar{x} + z_{a/2} \times \sigma] = 1\text{-}a$$

TABLE 1 Selected F-values and a-values for areas under a normal probability distribution curve (F-value = $z_{a/2}$)

$\mid z_{a/2} \mid$	1.00	1.28	1.64	1.96	2.00	3.00
a	.318	.200	.100	.050	.0456	.00270
a/2	.159	.100	.050	.025	.0228	.00135
1-a	.682	.800	.900	.950	.9544	.99730
(1-a)/2	.341	.400	.450	.475	.4772	.49865

This formula can be used to determine the confidence interval when we know the required probability and the standard deviation.

The situation of data stored in a GIS to be used for suitability mapping is slightly different. In this case, the confidence interval is provided, a value for the standard deviation is known, but what is required is the probability. The following two examples should serve to indicate how this can be achieved.

Example 1

Let us assume that for a land parcel to be very suitable for a crop the mean annual temperature must fall between 26°C and 34°C and that we have an estimated mean annual temperature ($\bar{x}$) of 30°C. Associated with temperature estimates is an SD of 2°C. In this case, the lower interval bound we are interested in is 26°C, which is ($\bar{x} - 2 \times$ SD). The upper interval bound is 34°C, which is ($\bar{x} + 2 \times$ SD). F-values of -2 and $+2$ therefore obtain. The value for ((1-a)/2) for F-values of $+/-2$ is 0.477; thus (1-a) is (0.477 + 0.477). In other words, there is a 95.4 percent probability (or 0.954 on the 0-1 range) that the land parcel is very suitable.

Example 2

Let us now consider the case where the mean annual temperature for a land parcel is 28°C. In this case, the lower interval bound is ($\bar{x} - 1 \times$ SD) and the upper interval bound is ($\bar{x} + 3 \times$ SD). The F-values are -1 and $+3$. Associated with these are ((1-a/2) values of 0.34 and 0.496, respectively; thus (1-a) is 0.836. It can be seen that there is only an 83.6 percent probability that the mean annual temperature of the land parcel is in the very suitable range.

LAND PARCELS

It should be noted at this stage that, although the term land parcel is used throughout this article, it may refer to a land parcel in the cadastral sense, or any other land unit such as a field, farm, cell, pixel, region, etc. Data processing is more straightforward if the land parcels are regularly-shaped cells; conceptualizing some of the data processing taking place in a GIS may also be easier if the land parcels are considered to be regularly-shaped cells. Planning decisions are made with regard to irregular shapes, such as fields, farms, or regions, however, so flexibility with regard to the shape of the land parcel must be retained.

At this point we can say that, outside the geo sciences dealing with topographic measurements, logic models are likely to be popular in geo information systems. It can also be said that although techniques exist (which will be described below) for determining a quality indicator in the form of the SD of the estimate when a mathematical model is being used, there are no recognized techniques for determining a quality indica-

tor when a logic model is being used. Furthermore, as implied above, financiers, planners, and others may be more comfortable with a quality indicator based on percentage probability.

In subsequent sections of this article, a procedure for generating percentage probability as a quality indicator when deriving suitability ratings using a logic model will be proposed, and procedures for deriving percentage probability values for the individual environmental parameters going into a GIS and contributing to a logic model used for suitability rating determination will be addressed.

ERROR PROPAGATION USING A MATHEMATICAL MODEL

In *Principles of Geographical Information Systems for Land Resources Assessment,* Burrough [5] discussed the application of an empirical equation called the "universal soil loss equation" (USLE). In this mathematical model, annual soil loss (A), in tonnes/hectare, is determined:

$$A = R \times K \times L \times S \times C \times P \qquad [1]$$

where:

R represents rainfall and runoff characteristics, which must be measured in centimeters, but is subsequently regarded as dimensionless;

K represents soil erodibility (a dimensionless factor);

L represents a topographic characteristic—the slope length, which must be measured in meters, but is subsequently regarded as dimensionless;

S represents slope in percentage terms;

C represents cultivation practices (a dimensionless ratio); and,

P represents soil protection measures (a dimensionless ratio).

Burrough provides values for these six soil characteristics based on Kenyan fieldwork, which also include standard deviations:

R = 297		σR =	72
K = 0.1		σK =	0.05
L = 2.13		σL =	0.045
S = 1.169		σS =	0.122
C = 0.5		σC =	0.15
P = 0.5		σP =	0.1

Thus using the USLE, the annual soil loss is:

A = 18.48 tonnes/hectare

Because this is a mathematical model, it can be used with conventional error propagation techniques to derive an SD for the most probable value for the estimated soil loss per year. The application of error propagation techniques to this model is outlined below.

σ A is unknown, but error propagation theory states that:

$$(\sigma A)^2 = (\frac{\delta A}{\delta R})^2 \times (\sigma R)^2 + (\frac{\delta A}{\delta K})^2 \times (\sigma K)^2 +$$
$$(\frac{\delta A}{\delta L})^2 \times (\sigma L)^2 + (\frac{\delta A}{\delta S})^2 \times (\sigma S)^2 +$$
$$(\frac{\delta A}{\delta C})^2 \times (\sigma C)^2 + (\frac{\delta A}{\delta P})^2 \times (\sigma P)^2 \quad [2]$$

where the terms $\frac{\delta A}{\delta R}, \frac{\delta A}{\delta K}, \frac{\delta A}{\delta L},$

etc, are partial differentials from equation [1]. (For example, the partial differential $\frac{\delta A}{\delta R}$ is $1 \times K \times L \times S \times C \times P$ and the partial differential $\frac{\delta A}{\delta K}$ is $R \times 1 \times L \times S \times C \times P$). To obtain a value for the SD of the most probable value of A, SDs for the parameters R, K, L, S, C, and P are used. Equation [2] acquires the form:

$(\sigma A)^2 = 0.00387 \times 5184 + 34180.711 \times 0.0025 + 75.339 \times 0.002 + 250.122 \times 0.014 + 1367.225 \times 0.0225 + 1367.224 \times 0.01$

thus $\sigma A = 12.4$.

Using the values given above and equation [1], it can be stated that the annual soil loss per hectare is 18.5 tonnes, with an SD of 12.4. The user of such a result, including this quality indicator, will realize that the error in all contributing variables was considered to be normally distributed and that there is therefore a 68.2 percent probability that the soil loss will lie between 6.1 tonnes/hectare and 30.9 tonnes/hectare, *ie*, 18.5 +/− 12.4 tonnes/hectare.

From the foregoing discussion, it can be demonstrated that the ability to estimate soil loss is constrained by the quality of contributing variables. Despite the USLE's attractive appearance of rigour, the possibly very large standard deviations in the contributing variables may diminish such an equation's usefulness unless we merely estimate that the derived value is more or less than a particular value. In the example calculated above, we can, for example, conclude—with an extremely high percentage probability—that the annual soil loss will be less than 75 tonnes/hectare. Thus it can be seen that mathematical models contribute to an eventual solution which will be found with a later application of a logic model. Cartographers will thus often find they need to consider error propagation in a mathematical model before proceeding to a logic model.

Furthermore, an empirical mathematical model such as the USLE may give rise to large systematic error. In this case, application of error propagation techniques is therefore not appropriate, and the only way of assessing the quality of derived results is experimentally—with the resulting quality indicator being only locally applicable.

DERIVING A PERCENTAGE PROBABILITY-BASED QUALITY INDICATOR IN A LOGIC MODEL

Let us consider, as a hypothetical example, a very simple logic model. Assume the success of a crop (crop X) is dependent on three independent variables—rainfall, elevation, and soil depth—and that each of these variables can be divided into three classes as follows:

Rainfall
R1 - more than 70 cm per year
R2 - 40 to 70 cm per year
R3 - less than 40 cm per year;

Elevation
E1 - less than 200 m asl
E2 - 200 m to 800 m asl
E3 - more than 800 m asl;

and,

Soil depth
S1 - more than 100 cm
S2 - 50 cm to 100 cm
S3 - less than 50 cm

Assume hypothetical experts have deemed that a land parcel will be very suitable for crop X if the annual rainfall on that land parcel is estimated to exceed 70 cm per year, the elevation is estimated to be less than 200 m above sea level, *and* the soil depth is estimated to exceed 100 cm. These same experts have deemed the land parcel is unsuitable for crop X if the annual rainfall is estimated to be less than 40 cm per year, the elevation is estimated to exceed 800 m above sea level, *or* the soil depth is estimated to be less than 50 cm. Any other situation renders the land parcel suitable as far as the experts are concerned. This represents one logic model. Clearly, others are possible and may be provided by alternative experts. This logic model may be more conveniently restated by using the classes referred to above: a land parcel is very suitable if its estimated values are classified as R1, E1, AND S1; a land parcel is unsuitable if its estimated values are classified as R3, E3 OR S3. In any other situation, the land parcel is suitable.

Let it be assumed that we already (somehow!) know the percentage probability that the variables in a given land parcel have values within the ranges represented by R1, R2, R3,....S3 and, furthermore, that rainfall, elevation and soil depth are uncorrelated. In this case, because the percentage probability that the land parcel is very suitable (VS) is based on the intersection of three independent events,

$$p(VS) = p(R1) \times p(E1) \times p(S1) \quad [3]$$

the percentage probability that the land parcel is unsuitable (U) is:

$$p(U) = p(R3); \text{ or } p(E3); \text{ or } p(S3)$$
whichever is the greatest $\quad [4]$

the percentage probability that the land parcel is

suitable (S) is the complement of the probabilities that it is very suitable and unsuitable:

$$p(S) = 1 - p(VS) - p(U) \qquad [5]$$

It can be seen from these equations (*[3, 4* and *5]*) that deriving all probabilities (VS, S, US) for each land parcel requires a percentage probability value for each characteristic in several classes, but how are these derived? Some consideration of this will be given in the following part of this article. As outlined above, an important contributor for determining the percentage probability value is SD; much subsequent discussion hinges on how to find this SD.

DERIVING PERCENTAGE PROBABILITY VALUES FOR CLASSES USED IN A LOGIC MODEL

All environmental information relating to a land parcel and stored in a GIS has been derived by point sampling or by defining boundaries. It is common for defined boundaries and their contents to be stored on maps, and maps as a source of information are considered below. The environmental information attached to a land parcel is derived by point sampling, often followed by interpolation (as may happen with elevations provided by a land surveyor or photogrammetrist, rainfall, soil depth, geochemistry, reflectances from a remote sensor, etc), or by deciding within which bounded area (polygon) a land parcel may fall (as may happen with administrative boundaries, elevation contours, soil zones, etc). The quality of the environmental information attached to the land parcel is thus a function of the quality of the original sample values and of the interpolation techniques used, or of the quality of the boundary line definition.

DERIVING PERCENTAGE PROBABILITY FOR VALUES OBTAINED FROM POINT SAMPLING

The example of crop X described in the preceding section relies on the consideration of three parameters whose values may have been estimated by point sampling (elevation, rainfall, soil depth) followed by interpolation. If information is available on the precision of the original point sample (its SD), the statistical facilities usually provided with interpolation programs (*eg*, DEM packages) can be used to generate an SD for interpolated values.

It is assumed that the estimated rainfall, elevation, and soil depth values for a hypothetical land parcel are:

Rainfall: 74 cm (with an SD of 8 cm)
Elevation: 142 m (with an SD of 5 m)
Soil depth: 164 cm (with an SD of 5 cm).

Applying the logic model outlined above, for the land parcel to be very suitable for crop X rainfall must be within a confidence interval with a lower value of

70 cm and an upper value of $+\infty$. The estimate of 74 cm is $(0.5 \times SD)$ above the lower interval bound. An F- value of 0.5 thus obtains. Associated with this lower F-value is a $((1- a)/2)$ value of 0.191. The estimate of 74 cm is $(\infty \times SD)$ below the upper interval bound. Associated with an F-value of ∞ is a $((1- a)/2)$ value of 0.500. The resulting (1-a) value is 0.691; or restating:
  - there is a 69.1 percent probability that the land parcel has very suitable rainfall (p(R1) = 0.691), also
  - the remaining probability can be allocated to R2, *ie*, p(R2) = 0.309, and thus
  - p(R3) = 0.000

This procedure can be repeated for elevation. The upper interval bound is 200 m, and the lower interval bound is 0 m. Elevation is 142 m with an SD of 5 m. The upper F-value is thus 11.6 and the lower F-value 28.4 Associated with the upper F-value is a $((1-a)/2)$ value of almost 0.500, and with the lower F-value also a $((1- a)/2)$ value of almost 0.500. Thus there is almost a 100 percent probability that the elevation value is very suitable, or:

p(E1) = 1.000
p(E2) = 0.000
p(E3) = 0.000

This can be continued for the soil depth, in which the following probabilities emerge:

p(S1) = 1 (or almost 1)
p(S2) = 0 (or almost 0)
p(S3) = 0 (or almost 0)

Thus using the formulae *[3]*, *[4]* and *[5]* we can conclude the following suitability ratings and probabilities for the land parcel under consideration:
  - The land parcel is very suitable for growing crop X with a percentage probability of 69.1 percent
  - The land parcel is suitable for growing crop X with a percentage probability of 30.9 percent; and
  - The land parcel is unsuitable for growing crop X with a percentage probability of almost 0 percent.

DERIVING PERCENTAGE PROBABILITY FOR VALUES OBTAINED FROM MAPS, WHEN A MAP SPECIFICATION EXISTS

The procedures discussed above for rainfall, elevation, and soil depth are suitable for variables which have been derived from point sampling and then have, perhaps, undergone interpolation to the land parcel. Much data going into GISs at the moment, however, come from maps on which boundary lines are presented. These boundaries may be administrative boundaries, soil class boundaries, lithologic boundaries, vegetation boundaries, contours, etc.

All boundaries as recorded on maps or in a data base have error associated with them. Even boundaries having a clear legal description and connecting points whose coordinates have been determined using the most elegant surveying and adjustment procedures, such as in a computational cadastre, acquire error. Although—for certain features such as complex soil classes having convoluted boundaries—the classification attached to a land parcel falling near the centre of a bounded area may be no less likely to be misclassified than one near the boundary [6], it can be intuited that the nearer a land parcel is to a boundary, the more likely it is that the parameter value assigned to it will be wrong.

Scientists in many disciplines are reluctant to say how precisely positioned are the boundaries related to their discipline; accuracy specifications, particularly for boundary maps of discontinuous themes, are not available. Several national topographic organizations are willing to make statements about the quality of their elevation contour lines, but such boundary lines, dealing with a continuous variable, are a special case because contour lines (eg, for elevation, depth, or gravity) will most probably—in a GIS—be treated as sources of sample points for interpolation to land parcels, and will only rarely be treated as the boundaries between (for example) land of different elevations.

Considering typically available accuracy specifications, it becomes clear that they can help only in situations which can be reduced to a point sampling problem. The United States Defense Mapping Agency (DMA), however, has in the past stated that "90 percent of all elevations determined from topographic maps shall have an accuracy with respect to true elevation of one-half contour interval or better" [7]. Most topographic map production organizations have straightforward quality control procedures to check that their map products reach these or similar standards. At the DMA (in 1971) this involved, for 1:25000 scale, using ground survey methods to determine the coordinates of at least 20 points along a 5 km profile, and checking these against those presented on the map [7]. Similar procedures may exist for organizations producing contour maps of other continuous variables, or could be adopted. It is unlikely that organizations producing maps of discontinuous variables (eg, soil classes, land use) will have such straightforward accuracy specifications, nor will it be possible for them to have such straightforward checking techniques.

But how can such a statement as that produced by DMA be used to get map quality information? Referring to Table 1, an F-value of 1.64 is associated with a normally distributed phenomenon with a (1-a) value of 0.90. Taking as an example a contour interval of 30 m and assuming the map has been produced to the above specification, then 90 percent of all check points have errors of 15 m or less. If it is assumed that heighting error is randomly and normally distributed on maps (as has been demonstrated in Malaysian maps [1]), then 15 m represents (1.64 × SD), or the SD associated with heights derived from a 30 m contour map is 9.1 m. The overall SD so obtained can be attached to any derived elevation.

DERIVING PERCENTAGE PROBABILITY FOR VALUES OBTAINED FROM BOUNDARY MAPS, NOT TAKING NEARNESS TO THE BOUNDARY INTO CONSIDERATION, WHEN NO MAP SPECIFICATION EXISTS

If values obtained from boundary lines are drafted to a general specification, only an overall SD can be derived for the characteristic, and this will be the same all over the map. This SD can be determined from the stated accuracy specification. Generally for thematic maps, other than those dealing with a continuous variable, a means has to be found for determining the quality of classification within the map. This problem was addressed by Hord and Brooner [12]. They resolved the "quality" problem into the three components:

(1) error in the boundaries' locations
(2) error in the map's geometry, and
(3) error in the classification

Their examination of the first component deals only with boundary error generated by the cartographic processes (manual drafting, raster plotters, etc). These are usually quite small and successfully "obscured" by the thickness of the boundary line, but perhaps only in the case of well-defined features on topographic maps will the total error in the boundary location be about the same as the line's width.

Error in the map's geometry can be conveniently determined at the control pointing stage in digitizing. Although quite straightforward, the application of this error in suitability mapping is not further considered in this article.

Hord and Brooner did considerable work in the determination of error in classification. They advocated field checking at sample locations to find the quality of the classification. Since their work in the mid-1970s, it has now become "respectable" to use some type of surrogate for field checking, such as 1:2000 scale photographs for checking a classification based on 1:10000 scale photographs [11], or satellite data for checking a very small-scale thematic map. Such checking can be used to determine whether land parcels are or are not correctly classified. If all land parcels in the sample are found to be correctly classified, an estimate of "correctness" of the classification of 100 percent probability can be produced. Of course, we can expect the checked sample to reveal a few errors, so lower probabilities are found. A formula for determining these probabilities is:

$$\bar{x} = 1/N \times \sum_{i=1}^{N} x_i \qquad [6]$$

where $\bar{x}$ is a probability estimate, based on a sample sized N, for the mean "correctness" (θ) for the whole population of land parcels, and where the percentage probability is:

$$\theta \times 100 = p \qquad [7]$$

114

and is estimated by:

$$\bar{x} \times 100 = p \qquad [8]$$

where $\bar{x}$ is the percentage probability estimate.

As the estimate, $\bar{x}$, of θ is a result of only sampling the population of land parcels, then a confidence level should be associated with this estimate. Aiming for a particular confidence level, for example 95 or 99.7 percent, it is possible to determine upper and lower values for the estimate of θ which will bound $\bar{x}$.

The "correctness" of the classification is being considered and is reduced to a binomial form (correct/ not correct). For a binomial distribution:

$$\sigma^2 = \theta(1-\theta) = p(1-p) \qquad [9]$$

Although reduced to a binomial form, the distribution of the "correctness" values of the classification can be approximated to a normal distribution [12, 2, 14]. Thus, for example aiming for a 99.7 percent confidence level for the percentage probability estimate (see equation [8]), F-values for the normal distribution (see Table 1) are used and give a band (or probability interval) extending ($3 \times$ SD, or $z_{a/2} \times$ SD) each side of the estimate, providing the upper and lower values for the band bounding the estimate. It is necessary to determine the size of this band or probability interval.

An approximate formula [2] is available for this determination:

$$\overline{\frac{\bar{x} - m\theta}{\sigma/\sqrt{N}}}^2 = (z_{a/2})^2 \qquad [10]$$

and this can be solved to give upper and lower values for θ. As an example, if 150 sample points were used and assuming that the initial estimate for the percentage probability of the correctness of the classification was 93 percent, and that the desired confidence level was 99.7 percent, then:

$$N = 150$$
$$\bar{x} = 0.93$$
$$z_{a/2} = 3.0$$
$$\sigma = \sqrt{(\theta(1-\theta))} \qquad \text{(from equation [9])}$$

thus:

$$N(x^2 - 2 \times x \times \theta + \theta^2) = (z_{a/2})^2 \times \theta(1-\theta) \qquad [11]$$

or

$$150(0.8649 - 2 \times 0.93 \times \theta + \theta^2) = 9.0 \times \theta - 9.0 \times \theta^2$$

and the resulting quadratric equation is:

$$159 \times \theta^2 - 288 \times \theta + 129.735 = 0 \qquad [12]$$

which can be solved to give the two values:

$$\theta = .97106$$
or,
$$\theta = .84025$$

Referring to equation [7], it can therefore be stated that the percentage probability that the classification at each land parcel is correct—at the 99.7 percent confidence level—is between 84.02 percent and 97.11 percent. Or to obtain a more useable statistic (there is a 99.7 percent confidence level that), the percentage probability value of the classification of 84 percent can be applied to all classifications in the area and used in processes such as applied in equations [3], [4], and [5].

DERIVING PERCENTAGE PROBABILITY FOR VALUES OBTAINED FROM BOUNDARY MAPS, TAKING NEARNESS TO THE BOUNDARY INTO CONSIDERATION, WHEN NO MAP SPECIFICATION EXISTS

As indicated above, scientists are reluctant to make statements about the quality of their boundaries. Blakemore [3] proposed the "epsilon" value as a measure of the quality of a boundary, but to derive an epsilon value for a boundary requires some knowledge of the possible spread in its definition which could be achieved by relevant professionals.

At ITC, we are fortunate in having, as course participants, mid-career professionals who are learning (for them) new techniques to be applied in their professions. Experienced soil or forest surveyors can be taught how to use aerial photos or satellite images for soil or forest surveys. In their courses, they have exercises in such techniques, and exercises which have been marked as correct by their mentors can, as a set, give an indication of spread for the types of boundaries concerned.

Figure 2 represents a superimposition of a set of five student exercises. It shows typical "spread" in boundary determinations. It can be seen that the greatest spread of any boundary is approximately 3 cm. A sample size of only five exercises is unlikely to be adequate, but the result achieved can at least be used in this discussion. Assuming:
- maximum spread is in practice the same as maximum error;
- a maximum spread of 3 cm; and
- the term "maximum" means 99.7 percent of all values,

then this value of 3 cm actually equals ($3 \times$ SD). It can be concluded that the SD of the boundary position is 1 cm, and also any land parcel more than 3 cm from the boundary can thus be deemed to be allocated to the correct bounded area.

Considering any land parcel less than 3 cm from the boundary, the percentage probability may be determinable from the actual distance of the land parcel from the boundary. For example, a point 2.00 cm from the boundary is $2 \times$ SD from the boundary and

115

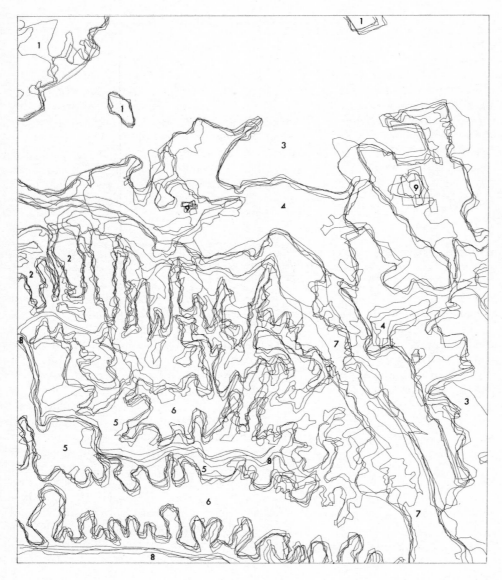

LEGEND

1 AGRICULTURE
2 RICE
3 WOODLAND
4 FALLOW BUSH
5 HIGH MANGROVE
6 LOW MANGROVE
7 GRASS MARSH
8 WATERCOURSE
9 VILLAGE

FIGURE 2 Superimposition of five student photo interpretation exercises in ITC's forestry group

the percentage probability that it is in the correct bounded area is 95.44 percent—which can be obtained from Table 1.

In this situation, where nearness to the boundary is significant, we are concerned with the intersection of two independent events (allocation to the correct polygon and .correct classification). That is, we have to determine the probability, p(C), that a land parcel has been correctly classified and has been allocated to the correct bounded area, when we know the percentage probability of correct classification, p(A), for example, 84.02 percent (*ie*, p(A) = 0.8402) and the percentage probability of correct assignment to a bounded area, p(B) (*ie*, p(B) = 0.9543).

$$p(C) = p(AB) = p(A) \times p(B) \qquad [13]$$

thus, $\quad p(C) = 0.8402 \times 0.9543 = 0.8018$

Such a percentage probability value could again be used in equations such as those found in *[3]*, *[4]*, and *[5]*.

APPLICATION

In the preceding sections, proposals for techniques for determining the percentage probability of a suitability classification were described when the suitability classification was derived using a logic model. These techniques used data attributed to land parcels which were:

- input to the system as sampled points, whose values were directly measured or determined from several parameters using a mathematical model (without systematic error), and were, perhaps, subsequently interpolated to land parcels;

- input to the system from maps showing boundary lines for which map specifications were available; or,

- input to the sytem from maps showing boundary lines for which no map specifications were available.

In this discussion, certain assumptions were made, including:

- that in every case some means of determining either an overall SD for the data set or a specific SD for data points was possible;

- that the best estimate for a sample point value was its most probable value which was the weighted mean of many observations, the observations had been acquired using acceptable measurement techniques, and error was normally distributed;

- that heighting error (and error in other continuous variables) was randomly and normally distributed on maps;

- that error in a map's geometry could be conveniently determined at the control pointing stage in digitizing;

- that the characteristics used in a logic model are uncorrelated;

- that maximum "spread" in boundary determinations performed by several competent professionals could be equated to the maximum error in boundary determination by one competent professional.

The procedures outlined in the preceding section represent the most general case for the transfer of thematic features from a map into a GIS. If the precision of the boundary is considered unimportant, $p(B)$ in equation [13] becomes 1 (or 100 percent); otherwise information on the boundary precision can be used. The information on boundary precision came from student exercises, and such information is not usually available. Even if it is available for one theme in one area, it is unlikely that the results could be transferred to other maps because thematic maps are usually produced for a small area and with a specific purpose in mind [6].

If information on boundary precision is not available, the procedure outlined for use in an absence of map specifications can be used. This makes no distinction between land parcels near boundaries and those in the centres of polygons, but requires checking at a large number of points either in the field or using surrogate products, such as large- scale aerial photos, to get percentage probability values.

It seems rather straightforward to use original survey sample values. This process was described above and a special case (ie, contour lines) was described in the subsequent section. Using instead a cartographic depiction as a source for the various environmental characteristics is cumbersome when trying to assess error. Unfortunately, however, data from original survey sample values are rarely available—sometimes because of expense. Many of the soil maps we now have evolved in an effort to reduce the expenses associated with gathering sample data. Soil maps may be generalized products showing the boundaries of "keys", such as soil colour, which the soil surveyor expected would coincide with particular physical and chemical properties of interest. But this generalization was also valuable because information on large numbers of soil characteristics could not be displayed or stored in any other way. Now we have information systems for the display, analysis, and storage of thematic data; maps as information systems are being replaced. Information systems should, as far as possible, take their data from direct measurements. Error analysis, along established error propagation lines, could then be used to obtain SDs for values for land parcels. Derived suitability values could then have associated with them a percentage probability value determined using the approach outlined for point sampling.

We began by suggesting that planners would be helped if a percentage probability, as a quality indicator, were attached to the suitability ratings they are given. This assumes that success will be achieved more often at land parcels which have been classified as very suitable with a high percentage probability, than at those which have been classified as very suitable but with a lower percentage probability. Investigation should, if possible, be carried out to determine whether this is true.

ACKNOWLEDGEMENTS

I am most grateful for the discussions I had with some of my colleagues during the preparation of this manuscript, and to Michael Weir, of ITC's forestry group, for the loan of the student exercises used to produce Figure 2.

REFERENCES

1 Abdul Kadir, B T. 1985. Investigation of a Photogrammetric Method to Evaluate Map Accuracy. MSc thesis, ITC Department of Photogrammetry.

2 Bhattacharyya, G K and R A Johnson. 1977. Statistical Concepts and Methods. John Wiley and Sons, Inc, New York.

3 Blakemore, M. 1983. Generalization and error in spatial data bases. Proc Autocarto VI, Ottawa.

4 Brady, N C. 1974. The Nature and Properties of Soils. Macmillan Publishing Company, New York.

5 Burrough, P A. 1986. Principles of Geographical Information Systems for Land Resources Assessment. Clarendon Press, Oxford.

6 Butler, B E. 1980. Soil Classification for Soil Survey. Clarendon Press, Oxford.

7 DMA. 1971. TOPOCOM Technical Manual TPC TM S-1,

United States Defense Mapping Agency.

8   Drummond, J E and P Stefanovič. 1980. Crop suitability mapping and associated problems. Proc Eurocarto III, Graz.

9   Ebdon, D. 1977. Statistics in Geography - a Practical Approach. Basil Blackwell, Oxford.

10 FAO. 1977. Soil and Land Resources Appraisal and Training Project - Philippines - Agusan River Basin, United Nations.

11 Fuller, R M. 1981. Aerial photographs as records of changing vegetation patterns. In: Ecological Records of Ground, Air, and Space - Institute of Terrestrial Ecology Symposium No 10, Natural Environment Research Council, Swindon.

12 Hord, R M and W Brooner. 1976. Land-use map accuracy criteria. Photogramm Eng and Remote Sensing, Vol XLII, No 5.

13 Rasdiani Sudibjo, E. 1984. Computer Assisted Cartography for Topographic and Thematic Maps. MSc thesis, ITC Department of Cartography.

14 Topping, J. 1972. Errors of Observation and Their Treatment. Chapman and Hall, London.

RESUME

Cet article aborde le problème tentant de fournir un paramètre de qualité lié à une information fournie par le système de géo-information (GIS). Une théorie de propagation des erreurs peut être appliquée à l'approche du modèle mathémétrique mais non à l'approche du modèle logique. Une procédure est proposée pour générer des paramètres de qualité pour des modèles logiques. On décrit des méthodes de dérivation de paramètres de qualité pour des données d'échantillons de points et des données de plans graphiques de limites. Leur importance est accentuée par le fait qu'elles donnent la possibilité d'estimer numériquement la précision des données utilisées comme entrées dans un GIS.

RESUMEN

Este articulo encara el problema de intentar proporcionar una cualidad asignada al parametro de la informacion suministrada por el sistema de informacion geografica (GIS). La teoria del error de propagacion se puede aplicar al modelo mathemetico pero no al modelo logico. Se propone un procedimiento para generar cualidades de parametros para modelos logicos. Se describen los metodos de deduccion de cualidades de parametros para los datos de prueba de punto y datos de mapas para los limites graficos. Se enfatiza la importancia de ser capaz de estimar numericamente la exactitud de los datos usados como entrada (input) al GIS.

MIKE WEHDE
Remote Sensing Institute
South Dakota State University
Brookings, SD 57007

Grid Cell Size in Relation to Errors in Maps and Inventories Produced by Computerized Map Processing

As cell size was allowed to increase, the accuracies of maps and inventories produced by computer processing decreased.

INTRODUCTION

INFORMATION obtainable through remote sensing techniques is becoming more useful to natural resource management and planning as it becomes integrated with other sources of information.[1] Computer processing is indespensible in storing, manipulating, retrieving, and displaying large quantities of diverse data.

Geographic information systems are used to produce inventories of resources in spatial units (acres, hectares, etc.) and to produce maps specifying the location of the resources. These products must be of known and reasonable accuracy to be acceptable to any decision process. Many articles in the field of remote sensing discuss accuracy of products resulting from process-

ABSTRACT: *Studies are reported which improve the understanding of the process of converting map data from a graphical representation to a computer compatible format. A uniformly shaped and spaced network of cells, a grid, may be used to determine the spatial characteristics of the map. Investigations were made into (1) a technique for characterizing the spatial nature of a map, (2) the effect of cell size and grid position on computer processing to produce inventory tables and new maps, and (3) the potential for modeling spatial cellularization.*

The frequency distribution of distances between boundary lines enclosing homogeneous map units was employed to characterize spatial characteristics of a map. The accuracies of maps and inventory tables produced by computer processing of a single map with different cell sizes and grid positions were determined. Grid position significantly affects accuracy when one isolated homogeneous map unit is processed; it is not significant in processing maps containing many such units. The importance decreases as the randomness of shape and size of the mapping units increases.

As cell size was allowed to increase, the accuracies of maps and inventories produced by computer processing decreased. Likewise, sample statistics (mean, mode, variance) of the interboundary distance distributions at each cell size were found to decrease systematically with the increases in cell size.

A mathematical modeling process was formulated to allow (1) estimation of the interboundary distance distribution of a map before cellularization and (2) prediction of mapping and inventory accuracies which might be achieved with different cell sizes. Two models were derived on the assumption that the quantities involved were one dimensional and were tested in comparison to experimentally observed accuracies. Although both models overestimated the errors at any particular cell size, the predictions were not erratic and the behavior of the models encourages further research into refinement of the models.

PHOTOGRAMMETRIC ENGINEERING AND REMOTE SENSING,
Vol. 48, No. 8, August 1982, pp. 1289-1298.

0099-1112/82/4808-1289$02.25/0
© 1982 American Society of Photogrammetry

ing to (1) identify mapping units by interpretation or machine assisted classification or (2) remove geometric distortions caused by the sensor. These articles are concerned with the correctness with which map units are identified and with the map geometry itself. This paper is concerned with computer processing of map data assuming the map is error free. The studies reported are concerned with understanding the effect of dividing maps into a grid of cells to allow computer processing.

Background

In order for an information system to be termed geographic and have the capability to generate maps, the data base must be designed to include spatial location information.[2] Location identifiers can be included in the data base by one of four techniques: external index, coordinate reference, arbitrary grid, and explicit boundary.[3] The latter two techniques maintain map boundary information in a form suitable for mapping. They are used in the two most common forms of geographic information systems, known as grid (cell) or line (polygon) systems, respectively.

A systematic comparison of the operating costs of cell and polygon system and of product accuracies was made by Smith.[5] He found that conversion of map data and typical analyses were eight to ten times more expensive with the polygon system although that system exhibits higher spatial accuracy. The faster, simpler cell systems are generally less expensive.[4,6]

A common criticism of cellular systems is that the gridding of the map for computer compatibility forces some selected grid cell size to be the lower limit on the spatial resolution. This makes cell size selection extremely important in creating a data base. Guidelines in the literature include utilizing the resolution of the source data,[7] selecting the smallest cell affordable in the operation of the system,[2] adjusting the size and shape of the cell to match the capabilities of the output device, e.g., rectangular to offset line printer aspect ratios,[8] and selecting a cell size small enough that the smallest mapping unit will be greater than 50 percent of any cell.[4,9]

A better understanding of the effect of cellularization on a map would be useful in (a) selecting the cell size for map conversion to computer format, (b) assigning cell size when converting from a polygon format to the cellular format, and (c) deciding on cell size changes during the course of map processing.

Cell size has significant effect on map accuracy. Nichols[10] reported a brief study of several cell sizes and soil map complexities and concluded that cellularization was too inaccurate. Hord[11] proposed a statistical model for evaluation of map accuracy, and Van Genderen[12] extended the application of the model to the problem of guarding against overconfidence. Note that the Hord-Van Genderen analyses require that the product map be in hand, and Tomlinson[13] reports that data preparation costs for a cellular system run four to five times the analysis costs. Hence, procedures for iterative digitization-evaluation-digitization are not appropriate to the problem.

Switzer[9] developed a map accuracy evaluation technique as a Boolean overlay of an "estimated" map and the "true" map with a two-level resultant map of matching and non-matching categories. Mathematical arguments and approximations allowed him to estimate the map accuracy from the "estimated" map alone. In the course of his analysis, he also justified square cells. His procedure, however, requires that the computer data be created before accuracy can be evaluated.

The Hord, Van Genderen, and Switzer procedures are useful in evaluation of product accuracies only in retrospect. The problem of cell size selection requires predictive capability. An estimation of product accuracy before data entry begins is necessary.

Mapping and Inventory Accuracies

The performance of a geographic information system may be measured by the accuracy of the products it produces, i.e., maps and tables (assuming that the map data in the data base are error-free). An experiment was conducted to seek a relationship between input map data characteristics and output mapping and tabulation accuracies with various cell sizes (for a detailed discussion consult Wehde[14]).

It should be emphasized that it is the cellularization of maps that is being studied, not a particular cellular information system. The information system employed for the study is described only to document the procedures by which the evaluation of cellularization took place.

The Area Resource Analysis System, AREAS, an information system developed at the Remote Sensing Institute, South Dakota State University.[15] AREAS provides the capability to change resolution (cell size), overlay maps, interpret maps, tabulate data sets, plot or record results and analyze data characteristics.[16]

A portion of a detailed soil survey map representing a two mile square area was selected as representative of a map with moderate polygon density yet diverse shapes and sizes of map units. To create a data base which could be employed as the "accurate" or "true" map standard, a very small cell size was selected. The cell size was also constrained to be a small subdivision of approximately 1, 4, and 16, ha (2.5, 10, and 40 acre) cells such that the study of increasing cell sizes would include these historically common sizes. A 0.007 ha (0.017 acre) cell met these requirements and resulted in a map data base of 384 cells per row in 384 rows. The original map and the base data set are shown in Figure 1.

Eleven additional data sets were created by adjusting the cell size. A grouping or aggregation of cells by integral multiples was employed, that is, pairs of cells in pairs of rows combined, groups of three cells in three rows combined, etc. In each succeeding case fewer cells of larger individual area represented the contents of the original map. Only those integral factors which evenly divide the 384 cell by 384 row map were utilized. This eliminated the situations of partial cells being created at the ends of rows or in the last row of the new map data set.

The integral factors employed to group cells into new map data sets were 2, 3, 4, 6, 8, 12, 16, 24, 32, 48, and 64. In the remaining figures and text these factors are termed "resolution numbers" or "resolutions" to maintain a context of spatial extent of the cell on the Earth's surface. The resolution numbers cited correspond to 0.028, 0.063, 0.112, 0.252, 0.448, 1.008, 1.792, 4.032, 7.168, 16.128, and 28.672 ha (0.069, 0.156, 0.278, 0.625, 1.111, 2.500, 4.444, 10.000, 17.778, 40.000, and 71.111 acres). The original, reference map data set and the eleven new map data sets created by the cell aggregation technique are mapped in Figure 2 by a film recording process.

The twelve data sets in the data base represent the same map cellularized at twelve different cell sizes. The AREAS information system was utilized to evaluate the accuracy of maps and inventories produced from each of the twelve data sets by the process shown in flow chart form in Figure 3. The process is shown for one resolution number and was repeated a total of eleven times. With the exception of the COMPARE step, all ovals in the flow chart signify an AREAS processing function, i.e., TABULATE, COMPOSITE, AGGREGATE, and INTERPRET.

The AGGREGATE function was written to relocate map boundarys among cells in a manner simulating larger cell sizes without actually reducing the number of cells or rows. This was necessary to allow the COMPOSITE function to overlay maps with a like number of cells for an analysis of combinations. Also, this kept all maps consistent in size (147,456 cells) to allow calculation of mapping error percentage based on the number of incorrectly assigned cells (INTERPRET as mismatch).

The percentage inventory error could not be obtained from the total cells tallied at each resolution because every inventory counted 147,456 cells. The inventory error had to be obtained from individual map theme or data categories present and then mathematically combined. Categories were inventoried as overabundant (+error) or too sparse (−error) on an individual basis. In total, these errors of commission and omission over

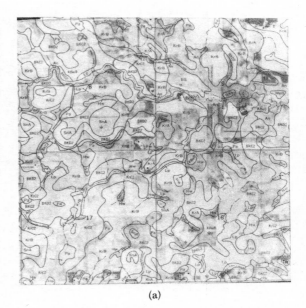

(a)

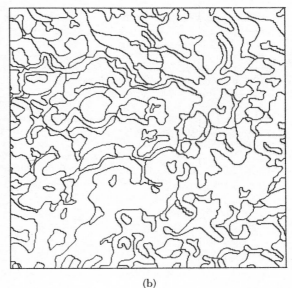

(b)

FIG. 1. A portion of the soil survey used for the experimental study. (a) the original data sheet and (b) the AREAS product at the 0.007 ha (0.0174 acre) cell size.

categories cancelled out. A root-mean-square error was calculated in order to avoid cancellation of errors. This, however, generated a value representing the average inventory error per map category rather than over the entire map. A root-sum-square calculation avoided the averaging over map categories to produce an inventory error for the map data set. The mapping and inventory errors determined for the twelve data sets are plotted in Figure 4. The relationship to cell size appears well

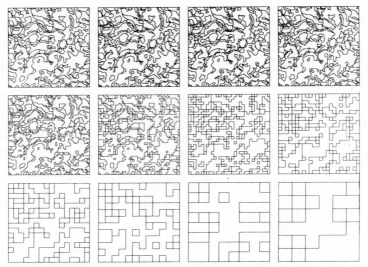

FIG. 2. The twelve maps analyzed for mapping accuracy with changing cell size. Cell sizes from top left to lower right are 0.007, 0.028, 0.063, 0.112, 0.252, 0.448, 1.008, 1.792, 4.032, 7.168, 16.128, and 28.672 hectares.

enough behaved to be modeled by curve fitting techniques, but no applicability beyond the present data would be achieved.

Since the cellularization process introduces error at the borders between homogeneous map regions, one might expect mapping error to depend on region size and perimeter (border length). The INTERPRET function was used to separate each of three map categories of differing abundance and complexity into map data sets. The mapping error evaluation diagrammed in Figure 3 was applied to these single-category data sets. Figure 5 shows the three data sets and the mapping error behavior with changing resolution number.

The BNE category with only one mapping unit demonstrates mapping error increasingly erratically with increasing resolution number until the cell size being created is too large for the unit to ever dominate the cell. From that cell size upward BNE is no longer represented at all in the map data set, hence mapping error is 100 percent. Smaller mapping units (polygons) within the category KRA or BKC2 would by themselves also exhibit a similar mapping error behavior with increasing resolution. The increasing abundance and size of polygons in these two categories allows increasing opportunity for more than one polygon to contribute to each of the larger cells being created and thereby provides opportunity for these categories to survive the larger cellular representation.

The mapping error graphed in Figure 4 for all map categories in the study, in comparison to the

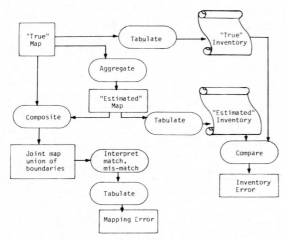

FIG. 3. The data processing diagram for experimental evaluation of errors with increasing cell sizes.

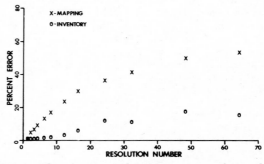

FIG. 4. The mapping and inventory errors observed as the cell dimension changes by a "resolution number," a factor representing magnitude of cell dimension increase.

mapping error graphs for the isolated map categories in Figure 5, demonstrates the averaging effect across map categories.

IMPORTANCE OF GRID POSITION

When a map unit is of a size approximately equal to that of the cells being used to represent the map, a wide variation in mapping error is possible depending on the position of the grid cells with respect to the map unit (Figure 6). The mapping errors graphed in Figure 4 were those resulting from positioning the sequence of increasingly larger grid cells in alignment with the row-one, cell-one position of the highest resolution data set.

When a cell dimension is doubled, the area increases by a factor of four. The resulting larger cells might be placed in alignment with either of two cell positions and either of two row positions in the smaller cell grid. In general, for a change of cell dimension by a factor of k there would be k^2 ways of positioning the new cell network over the old. The behavior of mapping error for individual mapping units, as shown in Figure 5, might be accounted for by the forced selection of a particular grid placement from the k^2 possibilities at each resolution k.

In order to evaluate this possibility, a mapping error analysis was conducted using a single circular polygon and a single rectangular polygon. At each new resolution, the mapping error for each possible orientation of the new grid cells in align-

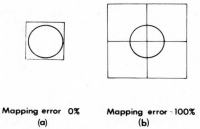

Mapping error 0%   Mapping error 100%
(a)   (b)

FIG. 6. The effect of grid position on mapping error for a single circular mapping unit. (a) the circle dominates the cell. (b) the circle does not dominate any of the cells.

ment with the old grid cells was recorded. The mapping error for the grid position aligned with row one and cell one was noted. Mapping error was averaged over all grid positions at each resolution.

A conceptually simple linear prediction of mapping errors for various resolutions was also defined for reference. The "linear model" simply predicts zero error at the reference or "true" map resolution of one, and 100 percent mapping error at whatever resolution exceeds twice the area of the mapping unit. Mapping error is linearly interpolated between these points. This model correctly represents the results of the dominant-theme cell encoding rule for simple closed mapping units or polygons of approximately square shape.

Figure 7 shows the mapping error for common point grid alignment, the mapping error averaged over all grid positions at each resolution, and the "linear model" of mapping error.

The results in Figure 7 do demonstrate that mapping error averaged over all possible grid positions is well behaved compared to the erratic behavior of mapping error observed in a sequence of aligned positions. An averaging principle is intuitively accepted: In the limit as k approaches infinity, the average mapping error for k positions of a grid over a single mapping unit is equivalent to the mapping error for k identical mapping units distributed throughout k positions on one grid. The requirement that $k \to \infty$ can be removed and equivalence can still be accurately maintained if the particular k positions of grids over mapping unit correspond one-to-one with the k positions of mapping units within one grid. Although many randomly placed identical mapping units are not often (if ever) found in practice, the averaging may be adequately approximated by a multiplicity of shapes, sizes, and placements of mapping units. In comparing the mapping errors for mapping units in Figure 5 to the overall map results in Figure 4, the averaging effect is apparent. Specifically, the conclusion is that grid positioning is not an important factor in dealing with a map (if the map has

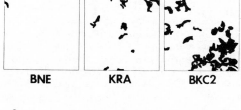

BNE   KRA   BKC2

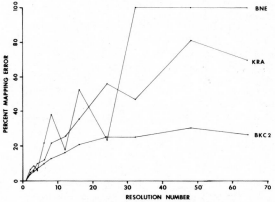

FIG. 5. Three selected soil mapping units and the mapping errors with changing cell dimension.

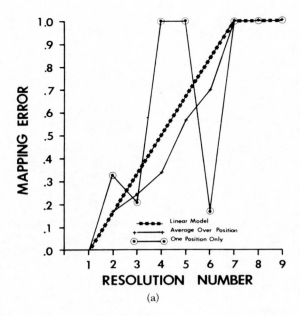

(a)

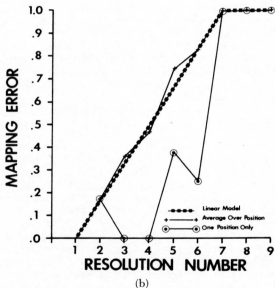

(b)

Fig. 7. Mapping errors for selected resolution numbers when the mapping unit is a simple closed figure. (a) a circle and (b) a rectangle.

a sufficient number of diversely shaped and/or diversely oriented mapping units to approximately satisfy the averaging principle).

Evaluating Map Structure

Map structure is the particular arrangement of various sizes and shapes of homogeneous polygons which uniquely defines the map itself. These same factors—size, shape, and arrangement of polygons—also determine the adequacy of any particular grid cell size for representing the map. A characterizing feature which simultaneously represents size, shape, and arrangement must be measurable or estimatible and offer a means for predicting cellular mapping error before maps are actually cellularized at any limiting resolution.

Since a map is a collection of spatially distributed boundary lines, the frequency distribution of distances between these lines must also uniquely represent the map. The distribution is continuous by nature since distance is a real numerical value subject to unit and scale influence. This continuous distribution is termed the inter-boundary distance distribution (IBD). Estimation of the distribution would be possible by randomly placing points and measuring boundary-to-boundary distance along random directions in order to enable construction of a relative frequency table. Since the intent of this paper is to use this IBD as a basis for modeling errors arising from finite cellularization of a map, the estimation of IBD and IBD statistics is restricted to the two orthogonal dimensions which would correspond to the rows of cells.

In a cellular information system, the interboundary distances are forced to take on some value which is an integer multiple of the cell dimension. In this sense the observation of interboundary spacing by analysis of a cellular information base becomes discrete. This discrete interboundary distance (in numbers of cells or multiples of the cell dimension) is called the span distribution.

A boundary analysis program of AREAS simultaneously scans along cells and down rows. The span distribution is generated for the "cells" dimension and the "rows" dimension and is combined for the map data set. Although a number of distribution forms are tabled, graphed, and statistically compared, the primary interest is the behavior of the map span distribution for each of the twelve data sets analyzed. These are shown in Figure 8.

The axis labeled "distance" indicates the number of cells between boundarys. Resolution one is represented in Figure 8a and is the best estimate of the map structure for the "true" map. At resolution six (Figure 8d) the most frequent boundary separation is one cell and certainly in the vicinity of resolutions 12 and 16 (Figures 8g and 8h) the data set becomes dominated by boundary separations of one cell. At approximately this point in the sequence of resolutions, the map structure is being overridden by the grid structure. The pattern of decreasing mean and variance corresponds to decreasing numbers of larger cells required to represent any particular interboundary distance.

The Span Distribution, as a two dimensional

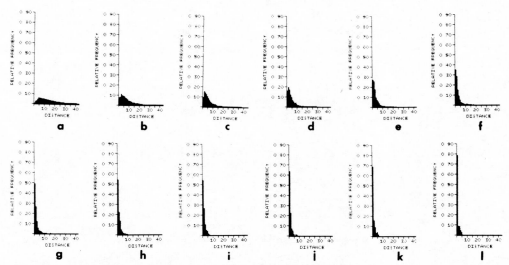

Fig. 8. Span distributions for the combined horizontal and vertical scans of the twelve study data sets in Figure 2. Resolution sequence is from upper left to lower right.

discrete estimate of the map IBD, behaves in a reasonable manner with changing resolution. Recall, however, that the distance in each graph of Figure 8 is measured in number of cells. The cell size for each graph is different. Taking the mean of each distribution times the actual cell size for that distribution allows the mean distance between boundaries to be plotted for each resolution number, as in Figure 9. The high degree of linearity may be accidental. If the placement of map boundaries were not affected by changing resolution, the increase to resolution k would decrease the mean of the distribution by $1/k$ and the graph would be of a constant, i.e., a horizontal line. The departure of the relationship from a horizontal line is an indication of the effect of cellularization on map structure. The slope and linearity of this relationship probably holds only for this particular map, and regression modeling is, therefore, not appropriate.

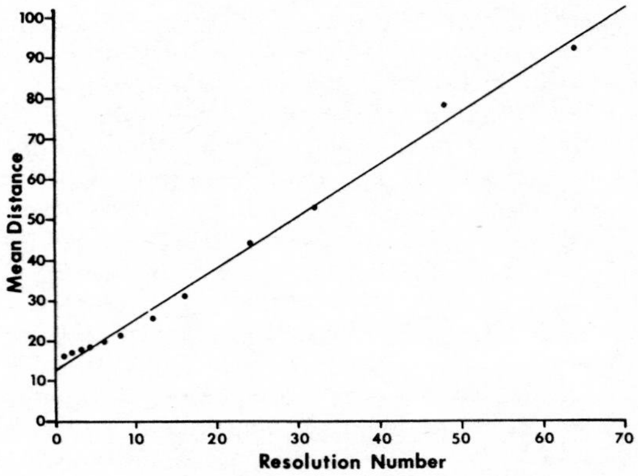

In establishing the existence of some relationship between mapping error and cell size (Figure 4) and between Mean Span Distance and cell size (Figure 9), the potential usefulness of the IBD concept for predicting mapping error is demonstrated. One should note that a statistic such as the mean is not unique to a map since more than one IBD could have the same mean. The IBD is unique to a map or set of maps. If more than one map has the same IBD, the application proposed in this paper would merely predict that all maps in the set would exhibit the same error behavior under various cellularizations.

PREDICTING MAPPING ERROR

For a particular map which is to be cellularized, the IBD can represent the important characteristics (polygon shape, size, and arrangement). A model of the errors arising from quantizing distances into discrete units of various sizes would then allow for mapping error prediction. The studies reported hereafter pursue derivation of the proper quantification process model. The span distribution of the "true" map data set was used as the best estimate of the IBD components in the cell and row dimensions.

A two-dimensional array was proposed as the form for the process model. Entries in the array would be some form of error estimate for the span and cell size combination corresponding to the row and column of the array. The process model array would operate on the span distribution to yield map error predictions for each cell size. A diagram of the array modeling technique proposed is shown in Figure 10.

Fig. 9. Mean Span Distance versus resolution number.

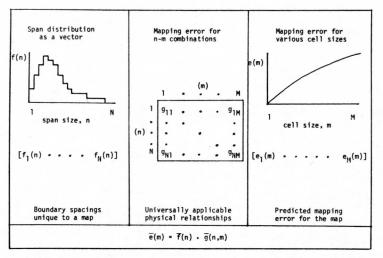

FIG. 10. Components of a simplified mathematical model between span distribution and mapping error vectors.

The span distribution, as an estimate of the interboundary distribution, is the input vector $f(n)$. The array model $g(n,m)$ represents the various choices of quantization sizes (cell sizes), expressed in m discrete-units, in relation to the span sizes, expressed in n discrete-units. The entries are estimates of mapping error for each combination. The vector-array product yields mapping error for various cell sizes, a vector $e(m)$.

THE POSITIONAL AVERAGE MODEL

The array $g(n,m)$ must contain entries representing error for the n-m combination of span size and cell size. Each entry represents a particular span, interboundary map distance. The averaging principle suggests averaging over possible grid positions (instead of averaging over mapping units).

This positional average array of mapping errors was the first experimental attempt at modeling. The derivation of the entries was based on observing enough combinations of spans and cell sizes to establish a predictive pattern in the entries. The simplifying assumptions made were that

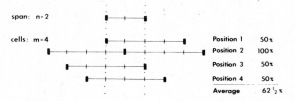

FIG. 11. An example of observing and calculating grid positional average mapping error in a one-dimensional sense.

(1) the observations were of one-dimensional quantities, i.e., line segments (spans) quantized into fixed increments (a cell dimension) with a dominant length coding rule, and (2) each line segment under study was considered isolated on a large homogeneous background.

The way the elements of the array were obtained is illustrated by Figure 11. Two map boundary lines are separated by a distance of 2 units (units are totally arbitrary—feet, inches, miles). If a cell size of 4 units is used to cellularize the map, there are four possible alignments on unit boundaries. Position 1 of Figure 11 shows a cell which lies 50 percent over a map category and 50 percent over a background category. In this tie situation the dominant length is randomly assigned. Hence 50 percent of the time the span is 100 percent incorrectly mapped (50 percent mapping error). In Position 2 the span is split between two 4-unit cells, with neither cell ever being categorized into the map unit. The span is never mapped correctly, hence the 100 percent mapping error. Positions 3 and 4 are equally likely variations of Position 1, and the overall average over possible grid positions is 62.5 percent.

Enough model array elements were calculated to observe a pattern, mathematically expressable for computer implementation. A computer program was written to generate the $g(n,m)$ to the dimensions required. The resulting $g(n,m)$ was applied in the manner outlined in Figure 10. The estimated mapping error from the model is compared to the experimental results in Figure 12.

Although the model strongly overestimates the experimentally observed mapping error, the trend and behavior of the model are encouraging. The

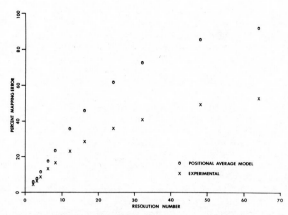

FIG. 12. The experimental mapping error and the mapping error predicted using the positional average model.

comparison is suggestive of a correct approach with perhaps oversimplifying assumptions.

THE SPAN ADJACENCY CORRECTED MODEL

The assumptions of the positional average model were oversimplifications of the two-dimensional cellularization process. Correction of the overestimation in Figure 12 was considered possible by (1) removing the span isolation assumption and (2) changing to a two-dimensional model (which would alter the array concept of Figure 10). The former approach was considered a more convenient alternative.

Removing the span isolation assumption allows spans to occur in sequence adjacent to one another. The occurrence of small spans adjacent to a span under study can alter the result of the dominant length rule used to decide what to do about fractional increments (Figure 13). The corrective effects are observed by comparing Figures 11 and 13.

The map background is homogeneous and of a category different from any of those represented by the adjacent spans. Only Positions 3 and 4 are repeated to demonstrate the corrective influence of the 1-unit span on mapping the 2-unit span with a 4-unit cell. Position 3 is now dominated by the 2-unit span and the category is correctly assigned to the cell, hence mapping error of 0 percent with respect to the 2-unit span.

The correction for span adjacencies requires consideration of the joint event, a span of n_1 units followed by a span of n_2 units. The occurrence of either span alone would be estimated by $f(n_1)$ or $f(n_2)$ from the span distribution. The estimation of relative frequencies for joint events can be simplified by assuming that spans occur independently of one another. Then the occurrence of a span of n units adjacent to any particular span of

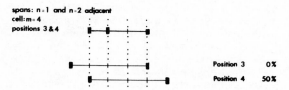

FIG. 13. An example of the positional average mapping error calculation when a one-unit span is known to be adjacent to the span of concern (compare to Figure 11).

interest can also be represented by $f(n)$, and joint event relative frequencies are simply the product of the relative frequencies of the components.

The procedure used in obtaining the Positional Average Array was repeated. The patterns of single and multiple adjacent spans which would reduce the mapping error by altering the dominance encoding step were recorded. The resulting correction array was of the same size $g(n,m)$. The entries were functions of $f(n)$, powers of $f(n)$ and multiple terms $f(n)$, $f(m)$, etc., corresponding to a single span, several equal spans, and a mixture of spans, respectively. The number of such terms in each entry mushroomed as n and m were increased. The span distribution of the reference map data set (highest resolution, smallest cell) had no span more frequent than 0.06; therefore, all higher power and cross product terms were dropped. The first-order, span-adjacency-correction array was generated by computer from the relationships of the first-order coefficients of $f(n)$.

The result of multiplying the array with $f(n)$ as in Figure 10 was a prediction of the correction to be applied to the positional average mapping error predictions. Figure 14 displays the experimental mapping error and the two array predictions for comparision. The span adjacent correction re-

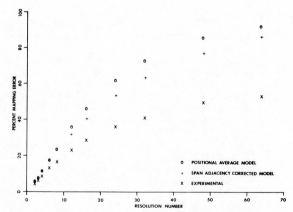

FIG. 14. The experimental mapping error and the predictions of the original and adjacency-corrected positional average models.

moved a part of the deviation of the prediction from the experimentally observed error.

Continued investigations appear warranted. The validity of span independence and first-order assumptions in the span-adjacency correction should be investigated. Also, the more difficult array model for two-dimensional cellularization by positional averaging and span-adjacency corrected positional averaging should be evaluated.

Conclusions

The studies reported have led to the following general conclusions:

- Grid positioning is not an important map accuracy consideration for maps but is a significant source of mapping error variation for individual map polygons.
- The interboundary distance distribution characterizes the size, shape, and arrangement of polygons in a resource map and is an appropriate input to a cellularization-process model.
- Interboundary distance distributions and mapping errors each relate to changing cell size in a well behaved fashion, making it likely that the process can be modeled.
- The grid positional average mapping error array is a significant and important component of a universal cellularization process model.
- The physical process of quantizing distances (cellularizing maps) can potentially be modeled accurately enough to allow prediction of mapping error for various cell size options and a particular map.

Acknowledgment

This work was made possible by grant NGL-42-003-007 from the National Aeronautics and Space Administration. A more detailed interim technical report SDSU-RSI-79-03 is available for reproduction and distribution.

References

1. Robert G. Reeves, Ed.-in-Chief. *Manual of Remote Sensing*, American Society of Photogrammetry, 1975.
2. D. Steiner and T. Stanhope. Data Base Development, Chapter 1 in *Geographical Data Handling*, prepared for UNESCO/IGU Second Symposium on Geographical Information Systems by the International Geographical Union Commission on Geographical Data Sensing and Processing, pp. 36-103, August, 1972.
3. R. F. Tomlinson, editor. *Geographical Data Handling*, published for UNESCO/IGU Second Symposium on Geographical Information Systems by the International Geographical Union Commission on Geographical Data Sensing and Processing, Ottawa, Ontario, Canada, 1972.
4. Richard L. Phillips. Computer Graphics in Urban and Environmental Systems, *Proceedings of IEEE*, Vol. 62, No. 4., pp. 437-452, April, 1974.
5. George Smith, Kris Van Gorkom, A. A. Dyer, *et al. Colorado Environmental Data Systems*, a final Report to the Colorado Department of Natural Resources by the College of Forestry and Natural Resources, Colorado State University, Fort Collins, Colorado, 1973.
6. D. Sinton. Introduction to Spatial Data Manipulation and Analysis, Chapter 8 in *Geographical Data Handling*, prepared for UNESCO/IGU Second Symposium on Geographical Information Systems by the International Geographical Union Commission on Geographical Data Sensing and Processing, p. 719, Aug. 1972.
7. Charles R. Meyers, Jr., Richard C. Durfee, and Thomas Tucker. *Computer Augmentation of Soil Survey Interpretation for Regional Planning Applications*, Oak Ridge National Laboratory Report ORNL-NSF-EP-67, April, 1974.
8. Phillip A. McDonald and Jerry D. Lent. MAPIT-A Computer Based Data Storage, Retrieval and Update System for the Wildland Manager, *Proceedings of the 38th Annual Meeting of American Society of Photogrammetry*, Washington, D.C., March 12-17, pp. 370-397, 1972.
9. P. Switzer. Estimation of the Accuracy of Qualitative Maps, *Display and Analysis of Spatial Data*, NATO Advanced Study Institute, edited by John C. Davis and Michael McCullagh, John Wiley and Sons, New York, 1975.
10. Joe D. Nichols. Characteristics of Computerized Soil Maps, *Soil Science Society of America Proceedings*, Vol. 39, pp. 927-932, 1975.
11. Michael R. Hord and William Brooner. Land-Use Map Accuracy Criteria, *Photogrammetric Engineering and Remote Sensing*, Vol. 42, No. 5, pp. 671-677, May 1976.
12. J. L. Van Genderen and B. F. Lock. Testing Land-Use Map Accuracy. *Photogrammetric Engineering and Remote Sensing*, Vol. 43, No. 9, pp. 1135-1137, Sept. 1977.
13. R. F. Tomlinson. Geo-Information Systems and the Use of Computers in Handling Land Use Information, in *Conference on Land Use Information and Classification*, sponsored by Department of Interior of U.S. Geological Survey and the National Aeronautics and Space Administration, Washington, D.C., June 28-30, 1971.
14. Michael E. Wehde. *Application Analysis of a Cellular Geographic Information System*, Masters Thesis, Electrical Engineering Dept., South Dakota State University, 1978.
15. Michael E. Wehde. *THE IMPLEMENTATION OF AREAS: Area Resource Analysis System*, Report SDSU-RSI-78-09, 1978.
16. Michael E. Wehde. *THE OPERATION OF AREAS: Area REsource Analysis System*, Report SDSU-RSI-78-10, 1978.

(Received 28 June 1979; revised and accepted 23 February 1982)

Accumulation of Thematic Map Errors in Digital Overlay Analysis

Jeffrey A. Newcomer *and* John Szajgin

ABSTRACT. The storage of digital resource information as geographically referenced data layers in computerized database systems has given managers access to a large variety of data that contribute to cost-effective management decisions. However, users of this information should be aware of potential shortcomings in using database techniques such as overlay analysis. Using fundamental concepts of conditional probability theory, the results of this investigation suggest that the accuracy of the map resulting from overlay analysis is determined by: 1) the number of map layers in the data, 2) the accuracy of these map layers, and 3) the coincidence of errors at the same position from several map layers. A method of calculating the upper and lower accuracy bounds of a composite overlay map is presented.
KEY WORDS: overlay analysis, conditional probability, intersection, mutually exclusive

Current requirements for increased planning in land resource management have created a need for greater quantities and varieties of basic resource information. This need is being met by more sophisticated data acquisition, processing, archiving, and retrieval methods. Foremost among these methods are computerized geographic information systems in which land resource data, such as soil type, land ownership, terrain information, land use, and land cover, are stored as geographically referenced "layers" or planes. These geographic information systems provide easily accessible resource information that can be conveniently manipulated and analyzed in developing management plans for large land areas.

Analysis procedures for digital map data can be placed in four major categories: 1) reclassifying map categories, 2) overlaying maps, 3) measuring map distance, and 4) characterizing cartographic neighborhoods (Tomlin and Berry 1979). This study investigates the technique of overlaying maps, which may be divided into arithmetic and boolean operations (Table 1). An example of an arithmetic operation is subtracting maps of land value at two separate dates to determine temporal changes in land value. An example of a boolean operation is the simultaneous evaluation of a vegetation map and an ownership map to determine the location of publicly owned forested areas. In this process, only those areas designated as forested and publicly owned would be portrayed as meeting the specified criteria on a final map or tabular summary. This can be thought of as a binary classification (two-class thematic map) in which all areas fall into one of two classes. That is, the areas either do or do not meet the specified criteria.

The verification and use of products from overlay analysis has generated concern about the accuracy of the information produced. In general, map errors can be categorized into two major types, positional errors and identification errors. Positional errors occur from inaccuracies in the horizontal placement of polygonal or cell boundaries. Identifica-

Jeffrey A. Newcomer is a Technical Programmer at Sun Exploration and Production Co., 4545 N. Fuller Drive, Irving, TX 75062. John Szajgin is Senior Applications Scientist at Technicolor Government Services, Inc., Bureau of Land Management Operations, Denver, CO 80225.

© 1984, American Congress on Surveying and Mapping, *The American Cartographer*, v. 11, no.1, p. 58–62; reprinted by permission.

Table 1. Types of overlay operations.

Arithmetic	Boolean
Addition	And
Subtraction	Or
Multiplication	Not
Division	Logical
Exponentiation	
Maximum (Ceiling)	
Minimum (Floor)	

tion errors occur from the mislabeling of areas of the various categories on the thematic map. Every thematic map overlay will have some combination of these two types of error. These errors can be the source of problems in overlay analysis when several thematic map layers are combined to address multiple decision criteria.

ERROR ACCUMULATION

Map accuracy can be related to the probability of finding what is portrayed on the map to be true in the field. Just as map accuracy can be related to probabilities, the combination of errors that can result from an overlay analysis can be demonstrated with fundamental formulas and definitions of probability theory. To start, let:

E_i = The occurrence of a correct value in layer i at a given location, where $i = 1, 2, \ldots, n$.

$Pr(E_i)$ = The probability that a correct value occurs in layer i at a given location, where $0 \leq Pr(E_i) \leq 1$.

E_i' = The occurrence of an incorrect value in layer i at a given location (expressed as the E_i complement).

$Pr(E_i')$ = The probability that an incorrect value occurs in layer i at a given location, where $0 \leq Pr(E_i') \leq 1$ and $Pr(E_i) + Pr(E_i') = 1$.

From this, an evaluation can be made to determine the accuracy of a composite map that results from overlay analysis.

For an overlay analysis involving two geographically registered sets of information, a correct value in the composite map results if there are correct values in each separate map layer in the data, that is, the events E_1 and E_2 have occurred. A wrong assignment would result from the combinations (E_1', E_2'), (E_1', E_2), or (E_1, E_2'). The probability of E_1 and E_2 occurring, or rather the intersection of the two events, is defined as $Pr(E_1 \cap E_2)$. Using Bayes Theorem (Hogg and Craig 1978), this can be expressed as

$$Pr(E_2|E_1) = \frac{Pr(E_1 \cap E_2)}{Pr(E_1)} \qquad (1)$$

or

$$Pr(E_1 \cap E_2) = Pr(E_1)Pr(E_2|E_1). \qquad (2)$$

In other words, the probability of the events E_1 (correct in layer 1) and E_2 (correct in layer 2) occurring together equals the probability of E_1 occurring multiplied by the probability of E_2 occurring given that E_1 has occurred— $Pr(E_2|E_1)$.

Similarly, the error combinations that can occur are expressed as

$$Pr(E_1' \cap E_2') = Pr(E_1')Pr(E_2'|E_1'), \qquad (3)$$

$$Pr(E_1' \cap E_2) = Pr(E_1')Pr(E_2|E_1'), \qquad (4)$$

and

$$Pr(E_1 \cap E_2') = Pr(E_1)Pr(E_2'|E_1). \qquad (5)$$

From these equations, another formulation for equation (2) is

$$Pr(E_1 \cap E_2) = 1 - [Pr(E_1' \cap E_2') + Pr(E_1' \cap E_2) + Pr(E_1 \cap E_2')]$$

or

$$Pr(E_1 \cap E_2) = 1 - [Pr(E_1')Pr(E_2'|E_1') + Pr(E_1')Pr(E_2|E_1') + Pr(E_1)Pr(E_2'|E_1)]. \qquad (6)$$

Equation 6 illustrates how the accuracy of a map product from overlay analysis can be, and usually is, less than the accuracy of the map layers that are put into the procedure due to the error accumulation that can result.

As the number of layers increases, the number of possible error combinations increases rapidly. For a three-layer example, each of the events E_1, E_2, and E_3 must occur simultaneously at a given location for a correct assignment to result.

The probability of the simultaneous, three-layer event can be expressed as

$$Pr(E_1 \cap E_2 \cap E_3) = Pr(E_1)Pr(E_2|E_1)$$
$$Pr(E_3|E_1 \cap E_2) \quad (7)$$

or

$$Pr(E_1 \cap E_2 \cap E_3) = 1 - [Pr(E_1' \cap E_2' \cap E_3')$$
$$+ Pr(E_1' \cap E_2 \cap E_3)$$
$$+ Pr(E_1' \cap E_2 \cap E_3')$$
$$+ Pr(E_1 \cap E_2' \cap E_3')$$
$$+ Pr(E_1' \cap E_2 \cap E_3)$$
$$+ Pr(E_1 \cap E_2' \cap E_3)$$
$$+ Pr(E_1 \cap E_2 \cap E_3')]. \quad (8)$$

Each of the seven terms on the right side of equation 8 can be expressed as a product of the conditional probabilities as in equation 7. This equation again shows the manner in which errors can accumulate and degrade the accuracy of the composite map. In general, the number of error combinations that can occur in an overlay procedure using n map layers is

$$\sum_{i=1}^{n} \frac{n(n-1)(n-2)\dots(1)}{[(i)(i-1)(i-2)\dots(1)]} \quad (9)$$
$$[(n-i)(n-i-1)(n-i-2)\dots(1)]$$

or

$$\sum_{i=1}^{n} \frac{n!}{i!(n-1)!} \quad (10)$$

ERROR ACCUMULATION EXAMPLES

A look at sample cases aids the understanding of this process. Assume two registered map layers of 100 elements (10 by 10) are to be used in an overlay analysis. The points in error in each layer are marked with an asterisk in Figure 1, and the points in error in the same position in both layers are circled. In this example, each layer contains 100 cells, five of which are in error. Thus, each layer is 95 percent correct, that is, $Pr(E_1) = Pr(E_2) = 0.95$ and $Pr(E_1') = Pr(E_2') = 0.05$). Note that four of the five error cells are coincident in the two layers. The probabilities that need to be determined are $Pr(E_1' \cap E_2')$, $Pr(E_1' \cap E_2')$, and $Pr(E_1 \cap E_2')$. From equation 2, the

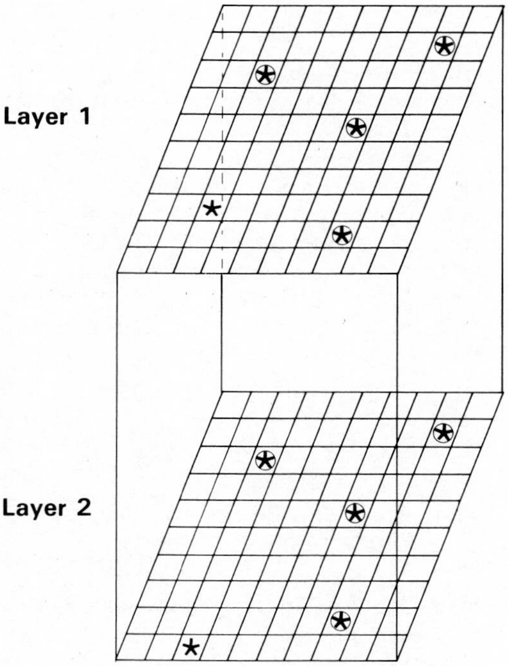

Figure 1. Coincident occurrence of several data-layer errors.

probabilities of the error combinations are

$$Pr(E_1' \cap E_2') = Pr(E_1')Pr(E_2'|E_1')$$
$$= (5/100)(4/5)$$
$$= 0.04, \quad (11)$$

$$Pr(E_1' \cap E_2) = Pr(E_1')Pr(E_2|E_1')$$
$$= (5/100)(1/5)$$
$$= 0.01, \quad (12)$$

and

$$Pr(E_1 \cap E_2') = Pr(E_1)Pr(E_2'|E_1)$$
$$= (95/100)(1/95)$$
$$= 0.01. \quad (13)$$

An example of the conditional probabilities needed here is $Pr(E_2'|E_1')$. The value of 4/5 (simultaneous errors) means that of the five cells in error in layer 1 (E_1'), four of the corresponding cells in layer 2 are also incorrect (E_2'). The two remaining conditional probabilities, $Pr(E_2|E_1') = 1/5$ and $Pr(E_2'|E_1) = 1/95$, are determined in the same manner. Using equation 6, the accuracy of the composite map that results is

$$Pr(E_1 \cap E_2) = 1 - [0.04 + 0.01 + 0.01]$$
$$= 0.94, \qquad (14)$$

which corresponds to a decrease in accuracy of 1 percent from each of the separate layers.

As a second example, assume an overlay analysis is conducted using two map layers in which each is again 95 percent correct (Figure 2). In this case, however, only one of the errors in layer 1 occurs in the same location as an error in layer 2. The probabilities of the error combinations that result are

$$Pr(E_1' \cap E_2') = Pr(E_1')Pr(E_2'|E_1')$$
$$= (5/100)\,(1/5) = 0.01, \quad (15)$$

$$Pr(E_1' \cap E_2) = Pr(E_1')Pr(E_2|E_1')$$
$$= (5/100)\,(4/5) = 0.04, \qquad (16)$$

and

$$Pr(E_1 \cap E_2') = Pr(E_1)Pr(E_2'|E_1)$$
$$= (95/100)\,(4/95) = 0.04. \quad (17)$$

The accuracy of the composite map in this example is

$$Pr(E_1 \cap E_2) = 1 - [0.01 + 0.04 + 0.04]$$
$$= 0.91. \qquad (18)$$

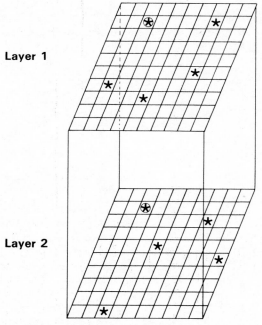

Layer 1

Layer 2

Figure 2. Coincident occurrence of few data-layer errors.

The difference in the accuracy of the two example composite maps is due to the orientation (simultaneous occurrence) of the errors. In the first example, the errors occurred at only six unique locations (Figure 1); however, in the second example, the errors occurred at nine unique locations (Figure 2). The occurrence of an error in layer 1 of the first example was highly associated with the occurrence of an error in layer 2 (4 out of 5). However, the occurrence of an error in layer 1 of example 2 has a low degree of association with occurrence of an error in layer 2 (1 out of 5). The result here is that as the errors in the map layers become less associated with the errors in all other map layers, the potential for error accumulation increases.

Assuming equal accuracies of the two map layers in the data, these examples illustrate that composite maps produced from layers with simultaneously occurring errors have higher accuracies than composite maps produced from layers with errors occurring at separate locations. In addition, the highest accuracy that can be expected is equal to the accuracy of the least accurate individual map layer. This only occurs when every erroneous point in each of the higher accuracy map layers being used occurs simultaneously with an erroneous point in the least accurate map layer. The lowest accuracy that can result is obtained when the errors in each map layer occur at unique point locations. In probabilistic terms, the errors are occurring at mutually exclusive, disjoint locations.

The second two-layer example illustrates this point if modified so that the circled error in layer 1 is moved one cell to the right. This causes all the errors to occur at separate locations as specified. From equation 2

$$Pr(E_1 \cap E_2) = (95/100)\,(90/95) = 0.90, \quad (19)$$

which gives a composite map accuracy less than any individual map layer and sets a lower bound on the accuracy of the composite map from this overlay analysis example. In general, the upper and lower bounds on the accuracy of a composite map from an overlay analysis of n layers are

$$\left[1 - \sum_{i=1}^{n} Pr(E'_i) \right] \leq \text{Composite Map Accuracy} \leq Min(Pr(E_i)), \quad (20)$$

where $i = 1, 2, \ldots, n$.

CONCLUSIONS

The results from the examples illustrate that the accuracy of a composite map from overlay analysis is generally less than the accuracy of the least accurate map layer used. A lower bound was also found to exist on the accuracy of a composite map produced from the overlay process. Because of error accumulation, even those maps that result from carefully planned overlaying will be less accurate than the map layers in the data. The routine compositing of map layers without regard to accuracy will then give rise to composite products of unknown integrity. However, lower accuracy and error accumulation does not imply that maps from overlay processes are not valuable. In fact, for many applications, the composite maps are valuable in that they portray the result of multiple-criteria decisions by marking areas that have potential for various land management practices.

The point to be made is that users of overlay techniques must assess the accuracies of the map layers being used in a given application. Knowledge of these accuracy values allows upper and lower bounds to be determined on the accuracy of a composite map. Users also need to utilize realistic decision criteria based on their knowledge of the resource to develop the most appropriate analysis procedure. By taking the decision criteria used and map-layer accuracies into account, managers can be assured of obtaining composite maps that are quite valuable when used with proper judgment.

REFERENCES

Hogg, R. V., and Craig, A. T. 1978. *Introduction to mathematical statistics*. New York: Macmillan.

McAlpine, J. R., and Cooke, B. G. 1971. Data reliability from map overlays. Paper presented at the 43rd Congress of Australian and New Zealand Associates for Advancement of Sciences, Division of Land Resources, Commonwealth Scientific Industrial Research Organization, Canberra.

Tomlin, C. D., and Berry, J. K. 1979. A mathematical structure for cartographic modeling in environmental analysis. *Proceedings*, ACSM 39th Annual Meeting, pp. 269–84.

Jeffrey A. Newcomer may now be reached at the following address: Section Manager, Science Applications Research, 4400 Forbes Blvd., Lanham, MD 20706.

John Szajgin may now be reached at the following address: Director of Onsite Operations, GeoSpatial Solutions, 882 E. Laurel Ave., Boulder, CO 80303.

SECTION 5
GIS Trends

Overview

In this section, recent developments and the future of GIS are discussed. Jackson and Mason, in the first article, describe an integrated geo-information system based on the merging of digital cartography and remote sensing, knowledge-based techniques, and advanced data structures. The potential of expert systems in GIS for resource management is explored by Robinson and others in the second article. The final article by Tomlinson Associates Ltd. provides an overview of GIS utilization in North America and technical and institutional prospects for GIS in the future.

Suggested Additional Reading

Published journals are the best source for information on trends in GIS. Examples include the *International Journal of Geographical Information Systems* published by Taylor and Francis, *Photogrammetric Engineering and Remote Sensing* published by the American Society for Photogrammetry and Remote Sensing, and the *American Cartographer* published by the American Congress on Surveying and Mapping.

Croswell, P.L. and S.R. Clark 1988. Trends in Automated Mapping and Geographic Information System Hardware. *Photogrammetric Engineering and Remote Sensing.* 54(11):1571–1576.

The development of integrated geo-information systems†

M. J. JACKSON

Laser-Scan Laboratories, Science Park, Milton Road, Cambridge, England

and D. C. MASON

Natural Environment Research Council Unit of Thematic Information Systems,
Department of Geography, University of Reading, Whiteknights,
Reading RG6 2AB, England

(Received 3 May 1985; in final form 27 November 1985)

Abstract. The paper provides a review of research and development in the field of integrated geo-information systems (IGIS) in general and at the Natural Environment Research Council's Thematic Information Service in particular. Advanced data-structures based on quadtrees are emphasized and tesseral addressing techniques and intelligent knowledge-based system approaches introduced. The paper first identifies how digital cartography and remote sensing have developed and merged within the IGIS concept. Recognition of the need for IGIS systems has stimulated research which increasingly is being based on hierarchical data models and incorporating knowledge-based programming techniques. A number of relevant research programmes are reviewed before presenting a more detailed description of NERC research into spatial data-base design, spatial addressing and arithmetic, knowledge-based image segmentation and classification and an expert system for map and graphic output. The paper concludes by identifying those key areas of IGIS design and development which require continued research priority.

1. Early developments: the cartographic data bank

Digital thematic cartography has passed through a number of distinct stages of development over the last 20 years (Jackson 1985). One of the first challenges was the development of computer hardware which would enable the digital capture of spatial data sets and their editing and reproduction in analogue map form through graphics and plotting facilities. Other early stages were aimed at the demonstration of the concept of digital map production. The products that were generated attempted to match conventional map design with the hope that, with continued development, the digital technology would lead to labour and cost savings. The considerable effort involved in data digitization and editing made both of these goals difficult to achieve and the benefits of speed and flexibility were instead increasingly emphasized. The next stage of development, therefore, concentrated on the mapping software and the design of a system which would produce a variety of map products quickly. The cartographic data bank began to be developed where data were no longer captured with a specific map in mind but rather with the aim of establishing an archive of data which could be displayed in many different combinations at different scales and map projections and with a variety of modes of presentation. The ECOBASE development at the

---

† An abbreviated version of this paper was first delivered at the RICS Symposium, Survey and Mapping '85, Reading University, 25–29 March 1985.

Experimental Cartography Unit† in the United Kingdom (Margerison 1976, Jackson 1982) is an early example of this development. To obtain the flexibility implied by the development of a generalized cartographic data bank placed constraints on its design which needed to avoid specific graphic conventions and to introduce flexible data structures. A 'vector' approach was developed by most organizations which recorded map data by encoding the left, right and on-line attributes of each feature perimeter. With minor variations this data structure has been employed in most of the more advanced digital mapping systems. Examples from the Thematic Information Service (TIS) are described by Jackson *et al.* (1983) and Harrison and Bell (1984).

Manipulation of spatially referenced data sets tends to be a computationally intensive operation. With the increasing power of computers and the greater availability of digital map data, a further stage of development occurred which moved the emphasis away from the generation of a conventional map product to the manipulation and analysis of the spatial data. As this stage developed, digital cartography began to constitute not only a means of graphical representation of spatial data but a powerful analytical tool. This latest change in emphasis is a fundamental one with a number of major implications to the role of digital cartography, its future development and the education of the future cartographer (Jackson 1984a). This wider role for digital cartography has only slowly been accepted but does now seem to be established. Thus, in a recent review of remote sensing and digital mapping by a subcommittee of the U.K. House of Lords Select Committee on Science and Technology (House of Lords 1983), a definition of digital mapping is given which includes, 'the use of any map data held in the computer, even though these are not reproduced in map form, and their integration with other data ... which are referenced by their geographical location'.

The role of digital cartography in the manipulation of spatial environmental data requires considerable structure within the cartographic data sets and processing environment together with powerful applications software. The cartographic data bank must become developed more fully as a geographic information system (GIS) with a well-structured integrated spatial data base. Furthermore, one can no longer restrict attention to vector map data sets but must also consider other related spatially referenced data. Two major additional forms of data become of concern, the first being point-referenced spatial statistics and the second, raster formatted remotely sensed data. It is to emphasize this important mixing of point, line, polygon and raster spatial data from several sources, and their incorporation within a single computational environment, that the prefix 'integrated' has been added to the more general GIS term in this paper.

2. The influence of remote sensing

Remotely sensed data and the growing disciplines of remote sensing and image analysis are now having a major impact on digital cartography even though the links between these subject areas are relatively recent (House of Lords 1983, Jackson 1984b). The stimulus has been generated by the rapidly growing remote-sensing community and many factors have contributed. These include the influx of new ideas from a wide variety of related disciplines, the availability of funding for space-related

† The Experimental Cartography Unit (ECU) was integrated into the U.K. Natural Environment Research Council's Thematic Information Service (TIS) in 1982. In 1985 the Cartographic Research Group of TIS moved to Reading University's Geography Department and was re-titled the NERC Unit of Thematic Information Systems.

activities, the access to and further development of sophisticated image processing hardware and software, the demands placed on digital cartography by remote-sensing programmes and the desire to utilize effectively the vast quantities of spatial data being collected by remote sensors. It is instructive to consider the growing influence of remote sensing in more detail and to consider how future developments might evolve. In the early days of environmental remote sensing there was a definite tendency (see, for example, Jackson 1985) to treat remote sensing in isolation, looking to the image data for total solutions to problems. The role of the map at this stage of development was limited to being a reference document and an analogue form was acceptable. Map data were increasingly used throughout the late 1970s and early 1980s not only for general reference purposes but for the selection of ground control points for image rectification, for identification, for geographical determination of training sites and for detection of land-cover change. Slowly, other map information came to be used for purposes of image segmentation and to assist image classification. Most of these links have been on a very *ad hoc* basis with digital data capture done on a piecemeal and project-by-project basis. Remote-sensing analysts are beginning to use digital map data and some digital cartographers are beginning to look to satellite data as a means of map update, but little genuine integration of the disciplines had occurred by 1983.

It was at about this time that the next major development occurred and this development was less of a 'technology push' and more of a 'user pull'. Useful applications of remotely sensed imagery were beginning to develop from the research programmes which had been undertaken in the 1970s. Once one started to use remote sensing as an operational tool then the image-derived data needed to be linked with other environmental data sets for further analysis. This placed further restraints on the image classification for it was now necessary to have one's field data and remotely sensed data compatible. At the same time it began to be realized that by feeding in to the image analysis system additional environmental data sets one could achieve more effective information extraction from the remotely sensed data. In placing the remotely sensed data properly into context, digital cartographers appreciated more fully the separate identity of the image analysis techniques from the satellite data. The image processing software and hardware provided an effective environment for manipulating spatial data in general. Sometimes the data sets might be dominated by satellite imagery but at other times they might consist largely, or even wholly, of more conventionally acquired data.

3.   The TIS experience

The developments which have occurred in the TIS demonstrate some of these trends. Until 1981 TIS had had only a peripheral interest in remote-sensing techniques. The Natural Environment Research Council (NERC), however, had recognized the potential of remote sensing to the environmental sciences and in 1981 invested in powerful image processing hardware and established an image analysis group. A number of the long-standing users of digital cartographic techniques welcomed the introduction of remote sensing and image processing facilities as a useful addition to the available sources of environmental data and of the tools for their analysis. With both facilities in the same building and operated by a single group there was a natural desire to exploit both facilities to maximize the benefit to the environmental sciences. The major obstacle of having vector-oriented hardware, software and data for the cartographic work and raster processing systems and data for the remotely sensed data, however, still existed. The availability of a multidisciplinary team of digital cartographers, remote-sensing specialists, mathematicians and computer and applications scientists provided an excellent base for the solution of such problems and within a short time scale a software interface existed between the two systems. This led to the ability to rasterize and overlay any of the encoded point, line or polygon cartographic data sets on any of the satellite remotely sensed data. A GIS was therefore developed and even though limited by its structure and functionality provided a significant step forward in the processing of spatial data.

The computational environment was now one which allowed the scientist to concentrate more on the problem in hand and exploit the spatial data sets appropriate to the problem on either system. In many cases this led to the integrated processing of vector digitized maps and remotely sensed data (Catlow *et al.* 1984, Baker and Drummond 1984). For some projects only satellite data was available (Baker and Allan 1983) while in others the data sets were all of a conventional form (Drummond *et al.* 1983). By 1984 the trend was moving away from narrow digital mapping and remote-sensing applications towards more generalized spatial environmental modelling. The *ad hoc* solution to integrating geographic data processing was quickly becoming restrictive as the size of data sets and complexity of processing requirements developed. A major programme of research and development was therefore defined and elements of it commenced with the ultimate objective of providing a fully integrated knowledge-based GIS.

4. IGIS—recognition of importance

While there are currently few groups in the U.K. with the necessary range of discipline expertise to carry out effective research in this area its importance is becoming widely recognized. A report to the Secretary of State for Education and Science by the U.K. Advisory Board for Research Councils in March 1984 (ABRC 1984, p. 18) highlights NERC's drive to enhance research in this area. The House of Lords Report (House of Lords 1983) likewise identified its importance and recommended that the Secretary of State for the Environment should appoint a Committee of Enquiry to report on the handling of geographic information in the U.K. (p. 73). This has now been established under the chairmanship of Lord Chorley, a member of the original House of Lords committee. The U.S. space organization, NASA, has sponsored GIS-related research (e.g. at Santa Barbara) and the European Space Agency has funded the European Association of Remote Sensing Laboratories (EARSeL) Working Group on Integrated Geo-information Systems to undertake a review study and assessment of research needs (EARSeL 1984).

While the developments quoted above have been from the NERC's TIS, similar developments have been occurring elsewhere. Thus, Scott (1984), describing joint work by the commercial company Dipix and the Surveys and Mapping Branch of the Federal Department of Energy, Mines and Resources, Canada, comments to the effect that it seems likely that integrated vector/raster image processing systems will become the rule rather than the exception over the next few years. Scott also predicts the further developments already occurring within TIS when he states: 'the previously divergent technologies of image analysis and graphics are now converging. Operational usage in digital mapping is the immediate goal, but these technologies promise ever broader applications in the future.'

5. Hierarchical data models and knowledge-based systems

The goal towards a genuinely integrated spatial processing environment for IGIS has thus been recognized and a start made on the necessary research and development. Early experiences suggest, however, that it will not be achieved merely by exchanging vector-based cartography for raster-based cartography and implementing an *ad hoc* integration with the raster satellite data. The spatial analyst/digital cartographer needs to re-examine fundamentally the approach to handling spatial data. Major changes in the current approaches to data structure, data addressing, data retrieval and user interaction with the GIS system will be necessary to produce a flexible IGIS with a tolerable response time to queries on the very large, multisource, data sets envisaged.

Two basic tools have been identified as being of fundamental importance in the development of such an IGIS. These are hierarchical data models and intelligent knowledge-based systems.

5.1. *Hierarchical data models*

The central problem in the design of an integrated spatial data base is the identification of a data structure suitable for holding all forms of spatial data (rasters, points, lines or polygons) in a compact form which allows efficient retrieval and processing. Recently, data models involving recursive regular decomposition of the plane have been proposed. Early research at TIS identified quadtrees and hextrees as potentially suitable structures, but other tessellations of the plane were also possible and there appeared to be no theoretical assessment of these options. An ongoing research effort involving TIS, NERC Computing Service and the Open University is therefore underway to find the best isohedral regular tessellation of the plane for the spatial processing of geographic data (this topic is discussed in §7.2).

The hierarchical data model adopted for the integrated spatial data base currently being developed at TIS is based on the quadtree. This involves a square tessellation and an amalgamation of four adjacent tiles to form the next higher-level tile. While the quadtree may not turn out to be the optimum choice on theoretical grounds, it has been extensively studied and a large body of algorithms exists to perform operations on quadtrees.

Figure 1 (*b*) shows the region quadtree representation of the binary image of figure 1 (*a*) (Samet 1984). Given a $2^n \times 2^n$ array of pixels, a quadtree is constructed by repeatedly subdividing the image into quadrants, subquadrants, etc., until blocks (possibly single pixels such as L in figure 1 (*a*)) are obtained consisting of a uniform property (in this case intensity). The process is represented by a tree in which the root node corresponds to the entire image, each non-terminal node has four sons, and the terminal nodes (leaves) correspond to uniform blocks. Both raster and vector data may be held in quadtree form, though different types of quadtree are necessary for holding point and line data.

The main advantages of hierarchical data models such as the quadtree are:

(1) All spatial relationships are implicitly present within the data model.
(2) They allow faster retrieval than the vector and raster models used in current GIS, because they employ an efficient 'divide-and-conquer' approach.
(3) They allow fast processing—an operation on a quadtree leaf need only be done once for the whole leaf rather than for every pixel in the leaf as with raster data.
(4) Raster data may often be held in a compacted form. The degree of compactness increases as the homogeneity of the image increases. While use of the quadtree

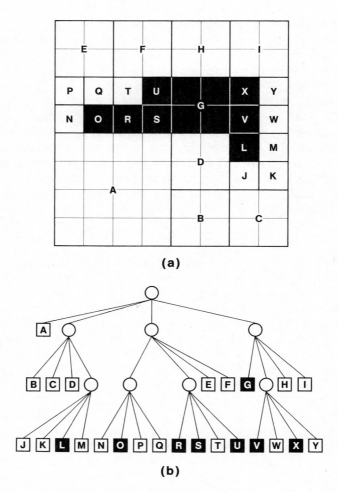

(a)

(b)

Figure 1. A binary image and its quadtree representation. (*a*) Binary image showing homogeneous blocks. (*b*) Quadtree.

may not result in the compaction of raw remotely sensed images, raster images held in a GIS are usually classified, or generalized, in some way.

(5) They allow the image to be viewed at different levels of generalization, e.g. the image may be viewed at low resolution for browsing, by looking only at higher levels in the tree. In cartographic terms the quadtree is a variable scale scheme based upon powers of 2. This is a major advantage for very large data bases.

It must be said that hierarchical models such as the quadtree have also been criticized on a number of counts (see, for example, McKeown 1983). One criticism is that their regular decomposition does not exploit explicitly the inherent structure in spatial organizations. Another is that if vector (line) data are converted to quadtree form (using, say, the edge quadtree described in §6), then resolution may be lost if the quadtree co-ordinate spacing is the same as that of the vector data. However, other data models also have their disadvantages (for criticisms of the relational model for handling spatial data see, for example, Marble and Peuquet (1983) and McKeown

142

(1983)). While the ideal spatial data structure may well involve a hierarchical component, its exact form is an open question.

5.2. *Knowledge-based systems*

The second major tool necessary to achieve the IGIS criteria required is likely to be the intelligent knowledge-based system (IKBS). The structure of knowledge-based systems can vary considerably but typically they consist of three major and separate components, (i) a set of rules (the knowledge base), (ii) a data base upon which these rules operate and (iii) a rule interpreter or inference engine. It is expected that their introduction will yield a system with data volume capacities, levels of computational efficiency and range of applications well beyond those which can be provided using traditional GIS approaches (Pequet 1984). Their introduction is probably farther off than the introduction of hierarchical data structures, as a large amount of research is necessary to develop systems capable, for instance, of achieving all the tasks set out below. However, separate systems could be developed incrementally and independently in the different areas.

Knowledge-based systems could be incorporated into a number of subsystems of the IGIS. Some of their main applications might be:

(1) To make the search procedure for complex geographic data bases more efficient by developing heuristic search techniques which limit the portion of the data base to be searched as early as possible in the search.

(2) To allow a capacity to learn, so that, for instance, a query that had been asked of the system on a previous occasion might be satisfied more rapidly the second time than the first; or so that the system might be able to adapt to new kinds of queries and applications.

(3) To assist in various ways in the provision of a 'friendly' user interface; for example, to enable a user scientist unfamiliar with the system to use the system in the most effective way, or to guide the non-cartographic user in appropriate map-design options at the output stage.

(4) To improve the classification of remotely sensed raster imagery prior to its inclusion in the data base. This could be done by incorporating into the classification procedure knowledge contained within data already resident in the data base, in the form of previous classifications, existing maps of the same region or domain knowledge derived from experts.

6. Review of related GIS work

In order to place the TIS IGIS research effort into a wider context, it is helpful to review the work of some other groups developing prototype GISs, especially those using either hierarchical data models or knowledge-based techniques. It should be emphasized that this is not intended as a complete review of current GIS work, merely a highlighting of those aspects which are relevant to the TIS programme.

Samet *et al.* (1984) have described an ongoing research effort to develop a GIS based on quadtrees. Quadtrees have traditionally been implemented as trees which require space for the pointers from a node to its sons. However, their approach is to employ a pointerless quadtree (Gargantini 1982) termed a linear quadtree. Only leaf nodes are stored in this structure, and a set of regions is treated as a collection of leaf nodes. Each leaf is represented by a hierarchical locational code specifying leaf position and size. The leaves are ordered by locational code in a sequence corresponding to a pre-order traversal of the tree. For example, if we consider figure 1 and store only the black leaves as in the original Gargantini scheme, the list would be L, O, R, S, U, G, V and X. Regions are represented by region quadtrees (Samet 1984) and for a leaf in a region quadtree the system stores the locational code and the leaf colour (this is a wider usage of the term 'linear quadtree' than proposed by Gargantini, who restricted it to black nodes only). Quadtree representations for point and line data have also been developed. In these the leaf colour is replaced by other information. Points are held using a PR quadtree (P for point and R for region) which associates data points with quadrants and in which no more than one data point may be stored in a quadtree leaf. The position of the point is stored in the leaf colour value. Linear data are held in a variant of the edge quadtree of Shneier (1981), in which a non-empty leaf contains exactly one straight-line segment which intersects two of the leaf's edges. The leaf colour value for line data becomes the intersection points. Special treatment is necessary at intersections and when a line only passes through one edge of a leaf. Use of the same basic data structure for regions, points and lines means that the same database kernel software can be used for all three data types. A key feature of their encoding scheme is that a pre-order traversal of the explicit tree produces nodes in ascending order of the addresses of the pixels at their lower-left corner. Given this linear ordering and the fact that the quadtrees stored (up to 4096 pixels × 4096 pixels) may contain as many as 30 000–40 000 leaf nodes, the quadtree is organized using a B-tree structure in which only that part of the tree currently being accessed need be held in core. This allows very large quadtrees to be stored and manipulated as virtual files.

Several data-base query functions have been implemented, including set operations (such as union and intersection), region property computations (area, perimeter) and map subset and windowing functions. A quadtree editor exists to allow the interactive construction and update of maps stored as quadtrees. The authors claim that a GIS based on quadtrees is feasible and that the potential advantage of using quadtrees rather than conventional data structures lies in the efficiency with which many types of query can be handled.

Mark and Lauzon (1984) are also experimenting with linear quadtrees to find if they are a suitable basis for efficient GISs. They are investigating the advantages, disadvantages and limitations of quadtrees relative to other GIS data structures. Attention is focusing on a two-dimensional run-encoded (2DRE) structure for representing a linear quadtree. A 2DRE structure results when a leaf-by-leaf linear quadtree is run-encoded on the attributes, that is, when only the locational code of the last leaf in a sequence of consecutive leaves of the same colour is stored, together with the colour. This is termed 2DRE because the sequence used for run-encoding is, in a sense, sorted or ordered in two dimensions at the same time. 2DRE quadtrees generally occupy less space than linear quadtrees so that for certain GIS processes (e.g. the important overlay procedure) the run-encoded files need never be decoded into individual quadtree leaves.

One of the largest and most advanced research programmes on knowledge-based GISs has been at the University of California at Santa Barbara (Peuquet 1984, Smith and Pazner 1984, Smith and Peuquet 1985). According to the authors, current GISs have consistently exhibited severe problems with response times and storage volumes for data sets of any significant size, as well as with rigidity and narrowness in their range of applications. The main reasons for this are that (i) geographic boundaries tend to be convoluted and irregular, and hence occupy much space, (ii) the digital data tend to be incomplete, imprecise and error-prone and (iii) spatial relationships tend to be fuzzy or application-specific, with the number of possible spatial relationships being very large. Current GISs also tend to be single format, i.e. not able to handle both raster and vector data.

The Santa Barbara work aims to develop a prototype knowledge-based GIS (KBGIS) which simultaneously addresses these problems. First, it employs a hierarchical data model (a quadtree) for the spatial data in the data base, because this allows greater flexibility and efficiency than the current raster or vector models. Secondly, it aims to use heuristic search procedures involving short cuts to increase data retrieval speed by eliminating large portions of the data base from consideration at an early stage of the search process. Heuristics utilize built-in data about the data, i.e. knowledge about how the various data elements are distributed within the data base. Thirdly, it employs a capacity to learn as a means of dealing with imprecision and uncertainty. This allows a system to provide answers in terms of 'confidence margins' and to adapt to new kinds of queries and applications. Learning is accomplished either inductively or deductively. Inductive learning is accomplished by the user supplying examples of new characteristics or relationships. Deductive learning is accomplished by the storage and accumulation of information gained from previous queries.

The dominant feature of the design of KBGIS is the dualism of functions relating to (i) spatial objects and (ii) the distribution of those objects in space. One function is designed to answer queries concerning objects at a specified location, the other to answer queries concerning the location of specified objects. A pair of hierarchical data models is used for these two functions, one spatially oriented and the other object oriented. The data are organized spatially in the form of pointerless quadtrees. The data are organized by object using and-or trees, which effectively record a taxonomy of data objects. The decision regarding whether to use the spatial or object portion of the data base in answering a particular query is made by a high-level query control system. This determines which path through the system is to be taken in answering each user query, using knowledge such as which combinations of lower-level tasks are needed to perform higher-level complex tasks and which is the shortest time path based on performance statistics from previous queries.

Another significant research programme employing knowledge-based techniques in GIS design is being undertaken in the Computer Science Department of Carnegie–Mellon University (McKeown 1983, 1984). This work has concentrated on the design of a knowledge-based system for map-guided image interpretation. The system developed, known as MAPS (Map Assisted Photointerpretation System), is a large integrated data-base system containing high-resolution photographs, digitized maps and other cartographic products together with detailed three-dimensional descriptions of man-made and natural features of the Washington, D.C., test area.

Other relevant research programmes are being undertaken by the Austrian Graz Institute for Image Processing and Computer Graphics (Ranzinger and Ranzinger 1984), the U.K. Kingston Polytechnic (Wilkinson and Fisher 1984) and, though mainly at the moment solely with map data, Davis and Nanninga (1984) at Australia's CSIRO Division of Water and Land Resources, Jungert et al. (1985) of the Swedish National Defence Research Institute and Bouille (1984) of the French Université Pierre et Marie Curie. Classified research is being undertaken by a number of industrial contractors and this may ultimately be expected to accelerate the commercial availability of such systems.

7. Components of the TIS IGIS research programme

Considerable attention has been paid within TIS to the conceptual design of an IGIS which will ultimately be knowledge based. The design aims to bring together spatial statistics, map data and image data into a single system where the environmental scientist can address his spatial problems without concerning himself with difficulties of data format differences. The structure of the current research model is described in detail by Jackson (1984c) and is not repeated here. However, four components of the TIS IGIS programme are described to highlight key aspects of the research. These components relate to:

(1) The spatial data-base design.
(2) The addressing and arithmetic operations on the spatial data.
(3) The role of IGIS in improving remotely sensed image classification.
(4) The increased need for, and opportunities offered by, expert systems in map and graphic output.

7.1. *Spatial data-base design*

The aim of this programme is to find out how best to manage large spatial data bases containing data integrated from a number of sources and for a variety of applications.

The proposed TIS solution employs a quadpyramid as the spatial data model. The quadpyramid is an extension of the quadtree in which information about lower nodes is held at higher nodes. For example, a high-level node might contain such information as the most frequent class in the subtree below that node. This feature of the quadpyramid should allow knowledge-based techniques to be incorporated into the data base. These would involve heuristic search procedures to avoid exhaustive search by eliminating large portions of the data base from consideration early in the search procedure, e.g. if it was recorded at a high-level node that no data about a particular object were stored at lower nodes, there would be no point in searching this subtree further. This is a simple example for illustration but the criteria may be more subtle and based on heuristics. Also, knowledge-based techniques could allow a capacity for the GIS to learn, as in the Santa Barbara approach.

Queries concerning the distribution of objects in space could be answered by searching the quadpyramid directly. However, for queries concerning objects (e.g. the attributes of points, lines and polygons) rather than their spatial distributions, a non-spatial data base existing side-by-side with the quadpyramid would be used to provide pointers to the relevant spatial data within the quadpyramid.

Development of the pyramid data base has begun with the building of a small experimental data base to be used as a test bed. The chief features initially aimed for will be to make the pyramid hold both vector and raster data and perform selected functions such as display and overlay. The key objectives at this stage are a better understanding of the data-base options, speed of processing, etc.

Following Samet *et al.* (1984) a memory management system for quadtrees has been constructed, which allows very large quadtrees to be stored and manipulated. The linearly ordered nodes are stored on disc and only that portion of the tree for which there is room resides in the core at any one time. A linked sequential B-tree whose index is a locational code is used to access the relevant virtual quadtree page. The section of the in-core quadtree first accessed is overwritten when a new section of the quadtree is brought in from disc, as there is a high chance that this will be spatially far from the region overwriting it.

Two main types of quadtree data structure are currently used, with higher nodes being kept for both. The normal quadtree form is a node list in pre-order tree traversal order, in which for each node a locational code, a node type and a colour are kept. The second form, used during quadtree building, is an implicit node list in post-order tree traversal order, for which only node type and colour are stored at each node, the node's locational code being calculable from its position in the list.

Currently, operations can be performed on region quadtrees generated from raster data. Several GIS analysis and display functions exist for these. Set operations, including intersection, union and exclusive OR are available, as well as region property computations such as area calculation. Display may be in either of two scan orders, a pre-order quadtree traversal, or a top-down breadth-first traversal, whereby all nodes at a given level are displayed, starting at the top level and working down. The latter gives a good demonstration of the generalization capability of the quadtree. A quadtree dilation algorithm allows buffering operations to be performed. Information regarding lower quadtree nodes may be stored at higher nodes, though currently this is limited to presence or absence of a class in the subtree below. A quadtree statistics utility generates various quadtree statistics, including size and spatial spectra (Chen 1984). Point data can also be handled, using a PR quadtree (Samet *et al.* 1984) — software currently exists for the generation of PR quadtrees, their display and set operations involving them. An English-like query language is used to access the data base.

The above project is based upon a data model involving a regular recursive decomposition of the image plane. An allied project having many similar features, but with the important proviso that it involves a data-driven irregular decomposition of the plane, is also being carried out for vector data (Robinson 1984). This should allow a useful comparison of the two approaches to be made. An irregular binary decomposition of the map plane is used to store and reference data. Summaries of the data ('directories') are held at all levels of the hierarchical binary tree which can be manipulated with relational data-base techniques. The individual data items (points, nodes, lines, polygons and text strings) are held on disc in a paged manner such that a page corresponds to a terminal node in the directory tree. Detailed information, such as co-ordinate strings, is held on separate pages. A novel feature of the design is the ability to display data at resolutions appropriate to the scale and output device. This is provided by holding for each line a binary strip tree which defines recursively the line in terms of rectangular envelopes.

7.2. *Spatial addressing and arithmetic*

The spatial data base requires an addressing system and arithmetic capable of allowing spatial operations (such as translation, rotation, etc.) to be performed on a hierarchical data structure. The NERC research team, in conjunction with the Open University, has achieved a significant advance in the development of tesseral addressing and arithmetic which allows the efficient manipulation of both vector and raster data irrespective of where within the plane the data lie (Bell *et al.* 1983, 1984, 1985 a, b, Bell and Diaz 1984.

Tesseral addressing is a replacement for Cartesian addressing in which a single address (a tesseral number) replaces the need for a two-entry x and y address. To understand the addressing system, consider the quadtree of figure 1 (*b*) representing the binary image of figure 1 (*a*). In this, single square pixels (e.g. K) may be considered part of squares (tiles) of side 2 pixels at the next highest level, and so on. A hierarchy of tiles is thus established covering the geographical space. All these tiles must be given an address which specifies the position and size of the tile. The addressing system used is shown in figure 2. Consider the image of figure 1 (*a*) superimposed on the upper-right quadrant of figure 2. Any point in the image can be addressed, e.g. pixel S has address 211. These tesseral addresses contain digits in the range 0 . . . 3. If we introduce a 'don't care' digit (say 4) we can address areas of greater size. For example, B in figure 1 has address 104, and A has address 044. For a $2^n \times 2^n$ image there are n tesseral address digits (leading zeros may be ignored) and these are numbers to the base 5. This part of the addressing system was first mentioned by Gargantini (1982). A TIS innovation was to extend the addressing system to cover the whole space, not just the top-right quadrant (Bell *et al.* 1983). This is illustrated by the numbers with dots in the other quadrants—these may be thought of as negative numbers. The dot indicates that the digit under it is repeated an infinite number of times to the left. This is reminiscent of the 2s complement method of holding a negative number which is used in computer arithmetic. It is essential that the tesseral arithmetic be extended to these other quadrants for IGIS purposes because one cannot pre-define the total area of interest or the need for spatial processing such as rotations and change of origin.

The quadtree may thus be represented in the data base in the form of a list of tesseral addresses, each held with the 'colour' of the image in the region covered by the address. The list may be sorted so that addresses which are close in space are also close in the list. The tesseral address may be employed in the data base in two roles, first, holding the tesseral address of data with the data to allow spatial processing and, secondly, allowing the creation of an index into the data base which uses the tesseral addresses as keys.

The tesseral arithmetic associated with the addressing system enables spatial

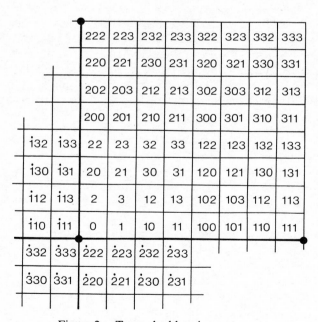

Figure 2.   Tesseral addressing system.

148

operations to be mapped directly into arithmetic operations. Tesseral arithmetic operations on the tile addresses induce transforms in the geographical image, such as translation, scaling and rotation, in a highly efficient manner. Tesseral addition produces a translation and tesseral multiplication a rotation and scaling, in the most general case. An example of addition is as follows. '0' assumes its usual role, i.e. $0 + a = a$. Thus if $0 + 1$ is thought of as a translation, it would cause a movement from the square with tesseral address 000 to that with tesseral address 001. This translation could be described as 'go one square in an easterly direction'. This translation would be the same at every point in the image, so that tesseral $003 + 1 = 012$, since 012 is 1 square east from 003. Similarly, the addition of 2 at any pixel would produce a translation of 1 square to the north, whilst the addition of 3 would produce a translation of 1 square to the north-east. The results of summing all possible pairs of digits 0, 1, 2 and 3 can be calculated and tabulated as in figure 3 (a). Sums with more than one digit follow the carry rule of normal arithmetic for taking the surplus digits on the left into the next column. In tesseral multiplication, 1 plays its usual role, so that $1 \times a = a$. Thus, if tesseral 001 is multiplied by 3, the result is 003 which is a rotation by 45° and a scaling by $\sqrt{2}$. Every tesseral address, considered as a vector from 000 to the square represented by that address, will be rotated anticlockwise by 45° and scaled by $\sqrt{2}$, when it is multiplied by tesseral 3. Similarly, every tesseral address, when multiplied by tesseral 2, will be rotated anticlockwise by 90° and scaled by 1. A tesseral multiplication table may be built up in a similar way to addition and this is shown in figure 3 (b). Long multiplication is analogous to that used in normal arithmetic, with the tesseral addition table used for the additions. It is also possible to do long division to any desired accuracy, which may include fractional tesseral numbers and addresses.

Tesseral arithmetic can be viewed as a more natural form of arithmetic for spatial operations than the normal Cartesian arithmetic. Just as ordinary arithmetic models the geometry of the number line, tesseral arithmetic models the geometry of the number plane. As ordinary arithmetic implies a translation back and forth along the number line, tesseral arithmetic translates on the plane, but in any direction. Similarly, as ordinary multiplication scales, and may change direction right or left of zero on the number line, so does tesseral scale, and may change the direction of the line between zero and the number multiplied, this time by any angle. Coupled with this is the advantage that tesseral algorithms are computationally more efficient than the equivalent Cartesian algorithms. Also, tesseral arithmetic and addressing can be implemented in hardware, which is important because it would allow fast parallel

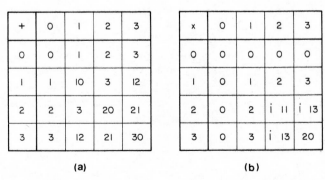

Figure 3.   Tesseral arithmetic: (a) addition and (b) multiplication tables.

spatial processing of images. Tesseral methods thus have many applications and far-reaching implications for the whole field of spatial data processing. Full details of tesseral methods are given in a series of publications by Bell, Diaz and Holroyd, who were primarily responsible for its development (Bell *et al.* 1983, 1984, 1985 a, b, Bell and Diaz 1984.

The discussions above have assumed a quadtree data model which is based on a square tessellation of the plane and an amalgamation of four adjacent pixels to form the next higher-level tile. However, similar tesseral addressing mechanisms and arithmetics exist for other tessellations, such as hexagonal and triangular, and for the various ways of amalgamating the basic 'atomic' tiles into hierarchies of larger composite 'molecular' tiles (Holroyd 1983). Future research will concentrate on the comparison of the various tessellations and amalgamations, the development of tesseral mathematics, the development of prototype tesseral hardware, tesseral data bases and the application of tesseral methods to higher dimensions.

7.3. *Knowledge-based image segmentation and classification*

Remote sensing, as well as being a major provider of data to an IGIS, will also be a major user of the same system in the goal of achieving maximum information extraction from imagery. The machine interpretation of remotely sensed images has relied in the past almost solely on the spectral analysis of pixels on an individual basis. The human photointerpreter, however, will use a much wider range of information. This will include context, such as where the region is in relation to others, topographic information, such as height, slope and aspect, information on previous classifications and general experience in the form of hunches or 'rules-of-thumb' which are not exact but which will usually lead to a correct interpretation. The latter will usually rely on the presence or absence of certain features or conditions or how they are associated in space (Jackson 1984 b). The IGIS provides a vital data base against which rules may be tested in association with the image classification. Moreover, the pyramidal structure of the data base can assist the classification procedure by assisting in the process of spatial and feature stratification. An example here might be that from a high level in the pyramid data base one could (i) stratify the whole image into major terrain categories such as lowland, steep hillsides and moorland, (ii) for each stratum, select out all non-water areas $\leqslant 5$ km from known (i.e. previously mapped) urban settlements and (iii) finally classify new urban developments within the defined search area on the basis of the spectral, contextual and textural characteristics of training areas chosen separately for each of the three strata.

A closely related project is in image segmentation. This project aims to increase classification accuracy by segmenting an image on the basis of information present in the KBGIS, in the form of previous classifications or existing maps of the same region. The output would be in a format directly suited to storage in a quad pyramid and would improve accuracy by enabling later classification to be carried out on a 'per field' basis (Mason *et al.* 1984, Cross and Mason 1985). A major extension of the segmentation project is currently proceeding as a joint industry/TIS/university collaboration under the U.K. ALVEY Information Technology programme.

7.4. *An expert system for IGIS graphics, map and image-generated products*

The potential users of the IGIS include planners, social and environmental scientists, the military, businessmen and engineers. In the past they have made some use of GIS but have been hindered by the limitations of hardware and the lack of integrated facilities of the type discussed in this paper. Both these limitations are being overcome and it will be increasingly practical to combine many spatial data sets at a time, view the result in colour, perhaps dynamically in 'movie' form and/or in perspective, to interactively edit, up-date, analyse, model and in other ways modify the resulting data and 'image' to meet specific needs. The results of these processes can also be passed increasingly easily to others either by transfer of the processed digital data or by photographing the screen or directing it to some form of colour raster plotter.

The users of these systems, however, are in general likely to be application specialists with limited cartographic or graphic-design expertise. It is also increasingly true that they are unlikely to have the direct support of professional cartographers. The potential power of the IGIS may therefore be partially lost through poor represent-ation of the spatial data and consequent loss of information in the interactive processing stages. Potentially as serious is the fact that in the presentation of spatial data for use by others the originator may fail to impart the information intended or even mislead.

The TIS IGIS research therefore includes a study aimed at reducing these dangers. The initial objective is to ensure that gross errors in data presentation do not occur and in the longer term, by gradual development, to have a system which will enable professional standard cartographic products to be produced by a non-expert. The approach being taken is to develop an 'expert' or knowledge-based system which will incorporate the design rules of cartographic experts. Rules may be progressively added to the system to improve its performance. Thus, one might start with simple rules such as do not use more than ten different colours on the same map or do not use blue to illustrate height of terrain which is not snow or ice covered. Later one might add in more exceptions based on the map context and detailed rules relating to colour balance between intraclass and interclass divisions, different levels of emphasis, different sizes of polygons, etc. Default options should then become more and more acceptable to specific users though user supremacy to select outcomes against the proffered system advice would also be possible. This study is being undertaken in association with Laser-Scan Laboratories Ltd., Glasgow University, Aberdeen University and King-ston Polytechnic. It is described in more detail by Robinson and Jackson (1985), who also define an expert system and look at how it differs from an ordinary (non-expert) programme.

8. Conclusion

The importance of IGIS and the related technologies has now been recognized by both the international science community and by national and international, govern-ment and commercial user organizations. Fundamental improvements in GIS design are required, however, if the full benefits of remote sensing and the advances in information technology are to be exploited. The paper has considered some of these issues and described selected elements of the TIS's research programme. Further fundamental research is required and care is needed to ensure the enthusiasm for the IGIS concept does not lead to implementations based on an inadequate understanding of the problems which arise from spatial integration and spatial data-base operations with very large, multisource data sets.

This paper has emphasized the development of a quadpyramidal data base employing new, highly efficient, data-addressing techniques for efficient spatial query of very large and multidata type data bases. It has also suggested the importance of long-term developments in intelligent knowledge-based (or expert) systems techniques. Topics of priority now include the implementation of such a system in demonstrator form for experimental purposes. Associated with the evaluation of the demonstrator system should be the development of a user-friendly man–machine interface to ensure effective exploitation of the IGIS by non-experts. This is also likely to incorporate IKBS techniques. These developments in turn should open up new opportunities in the field of remote sensing by improving the accuracy of image classification through the use of IGIS supported contextual and IKBS-based classification procedures. Finally, the relevance of new hardware technology, particularly developments in the design of parallel architectures, dedicated data-base machines and graphics workstations, and of algorithm development in the fields of pattern recognition and computer vision, need to be more fully evaluated and incorporated into GIS research programmes.

Acknowledgments

The paper is published with the permission of the Director, Natural Environment Research Council Scientific Services, U.K.

References

ABRC (ADVISORY BOARD FOR THE RESEARCH COUNCILS), 1984, Scientific opportunities and the science budget 1983. Report to the Secretary of State for Education and Science from the ABRC, March 1984, 65 pp.

BAKER, J. R., ALLAN, T. D., and LINTON, R. H. W. (editors), 1983, Digital analysis of Seasat SAR imagery over the JASIN area. In *Methods of Display of Ocean Survey Data* (Swindon: NERC), pp. 86–103.

BAKER, J. R., and DRUMMOND, J. E., 1984, Environmental monitoring and map revision using integrated Landsat and digital cartographic data sets. *Int. Arch. Photogramm.*, **25**, 25.

BELL, S. B. M., and DIAZ, B. M. (editors), 1984, *Proceedings of the First Tesseral Workshop*, Swindon, 13–14 August 1984.

BELL, S. B. M., DIAZ, B. M., and HOLROYD, F. C., 1984, Tesseral image processing. *Proceedings of the Remote Sensing Society Workshop on Digital Processing in Remote Sensing*, London, 28 March 1984.

BELL, S. B. M., DIAZ, B. M., and HOLROYD, F. C., 1985 a, Tesseral addressing and tabular arithmetic of quadtrees. Technical Report NERC/TIS, Holbrook House, Station Road, Swindon.

BELL, S. B. M., DIAZ, B. M., and HOLROYD, F. C., 1985 b, Tesseral addressing and arithmetic—an introduction. Technical Report, NERC/TIS, Swindon.

BELL, S. B. M., DIAZ, B. M., HOLROYD, F. C., and JACKSON, M. J., 1983, Spatially referenced methods of processing raster and vector data. *Image Vision Comput.*, **1**, 211.

BOUILLE, F., 1984, Architecture of a geographic structured expert system. *Proceedings of the International Symposium on Spatial Data Handling*, Zurich, August 1984, Vol. 2, pp. 520–543.

CATLOW, D. R., PARSELL, R. J., and WYATT, B. K., 1984, The integrated use of digital cartographic data and remotely sensed imagery. *Proceedings of the EARSeL/ESA Symposium on Integrative Approaches in Remote Sensing*, Guildford, 8–11 April 1984 (ESA SP-214), pp. 41–46.

CHEN, Z.-T., 1984, Quadtree spatial spectrum: its generation and application. *Proceedings of the International Symposium on Spatial Data Handling*, Zurich, August 1984, pp. 218–237.

CROSS, A., and MASON, D. C., 1985, Segmentation of remotely sensed images by a split and merge process. *Proceedings of the International RSS/CERMA Conference on Advanced Technology for Monitoring and Processing Global Environmental Data*, London, pp. 177–189.

DAVIS, J. R., and NANNINGA, P. M., 1984, Geomycin: towards a geographic expert system for resource management. CSIRO Division of Water and Land Resources Report, Canberra, ACT 2601, 32 pp.

DRUMMOND, J. E., GREEN, P., JACKSON, M. J., and PLANT, J., 1983, A regional raster data study at the Experimental Cartography Unit: the application of interactive raster graphics to environmental modelling. *Proceedings of the Annual Conference of the American Congress of Surveying and Mapping*, Washington, D.C., pp. 294–401.

EARSeL (EUROPEAN ASSOCIATION OF REMOTE SENSING LABORATORIES), 1984, Study on analysis of integrated geo-information systems. WG13, Annex A, Technical Proposal for ESOC RGQ13-4976/84/0/JS, 30 pp. Available from the Working Group Chairman, Dr. M. J. Jackson, NERC, Swindon.

GARGANTINI, L., 1982, An effective way to represent quadtrees. Commun. Ass. Comput. Mach., **25,** 905.

HARRISON, A. R., and BELL, S. B. M., 1984, Recent database developments at the Experimental Cartography Unit. Technical Report, NERC/TIS, Holbrook House, Station Road, Swindon, 17 pp.

HOLROYD, F. C., 1983, The geometry of tiling hierarchies. *Ars Combinatoria*, Vol 16-B, pp. 211–244.

HOUSE OF LORDS, 1983, *First Report by the Select Committee on Science and Technology on Remote Sensing and Digital Mapping*, Vol. 1 (London: HMSO), 88 pp.

JACKSON, M. J., 1982, Automated cartography at the Experimental Cartography Unit, NERC. In *Computers in Cartography*, edited by D. Rhind and T. Adams. British Cartographic Society Special Publication No. 2 (London: British Computer Society), pp. 132–142.

JACKSON, M. J., 1984 a, New technology and its present and likely future impact on cartography. In *Cartographic Education for the Future*, edited by G. R. P. Lawrence. British Cartographic Society Special Publication No. 3 (London: British Computer Society), pp. 59–64.

JACKSON, M. J., 1984 b, A methodology for an integrated spatial data model and its application to image classification. *Proceedings of the EARSeL/ESA Symposium on Integrative Approaches in Remote Sensing*, Guildford, 8–11 April 1984 (ESA SP-214), pp. 269–273.

JACKSON, M. J., 1984 c, Image analysis and digital cartography within the NERC Thematic Information Service: the development of a unified spatial model. *Photogramm. Rec.,* **11,** 383.

JACKSON, M. J., 1985, Digital cartography, image analysis and remote sensing: a move towards an integrated approach. *Interdiscipl. Sci. Rev. J.,* **10,** 8 pages.

JACKSON, M. J., BELL, S. B. M., and DIAZ, B., 1983, Geographical data base developments in NERC Scientific Services. *Cartographica,* **20,** 55.

JUNGERT, E., FRANSSON, J., OLSSEN, L., TOLLER, E., BORGEFORS, G., LINDGREN, T., and ROLDAN-PRADO, R., 1985, Vega—a geographical information system. FOA Report D 30367-E1, National Defence Research Institute (FOA), Linköping.

McKEOWN, D. M., 1983, MAPS: the organisation of a spatial database system using imagery, terrain and map data. *Proceedings of the DARPA Image Understanding Workshop*, Crystal City, Virginia, June, pp. 105–127.

McKEOWN, D. M., 1984, Knowledge-based aerial photo-interpretation. *Photogrammetria,* **39,** 91.

MARBLE, D. F., and PEUQUET, D. J. (editors), 1983, *Geographic Information Systems and Remote Sensing, Manual of Remote Sensing* (American Society of Photogrammetry), Virginia, pp. 932–958.

MARGERISON, R. A., 1976, *Computers and the Renaissance of Cartography*. Natural Environment Research Council, Swindon, 20 pp.

MARK, D. M., and LAUZON, J. P., 1984, Linear quadtrees for geographic information systems. *Proceedings of the International Symposium on Spatial Data Handling*, Zurich, August 1984, Vol. 2, pp. 412–439.

MASON, D. C., BELL, S. B. M., CROSS, A., FARNELL, E., and HARRISON, A., 1984, Image segmentation using a hierarchical data structure. *Proceedings of the First Tesseral Workshop*, Swindon, 13–14 August 1984.

PEUQUET, D. J., 1984, Data structures for a knowledge-based geographic information system. *Proceedings of the International Symposium on Spatial Data Handling*, Zurich, August 1984, Vol. 2, pp. 372–391.

RANZINGER, H., and RANZINGER, M., 1984, A geo-information expert system for synergetic use of map and image data. *Proceedings of the EARSeL/ESA Symposium on Integrative Approaches in Remote Sensing*, Guildford, 8–11 April 1984 (ESA SP-214), pp. 263–268.

ROBINSON, G. J., 1984, COMPASS II outline specification. NERC/TIS Internal Report, Swindon, 8 pp.

ROBINSON, G. J., and JACKSON, M. J., 1985, Expert systems in map design. *Proceedings, Auto-Carto 7*, Washington, D.C., 11–14 March 1985, pp. 430–439.

SAMET, H., 1984, The quadtree and related hierarchical data structures. *ACM Comput. Surv.*, **16**, 187.

SAMET, H., ROSENFELD, A., and SHAFFER, C. A., 1984, Use of hierarchical data structures in geographic information systems. *Proceedings of the International Symposium on Spatial Data Handling*, Zurich, August 1984, Vol. 1, pp. 392–403.

SCOTT, A. J., 1984, Integrating raster and vector systems for mapping. *Comput. Graph. World*, February 1984, pp. 63–66.

SHNEIER, M., 1981, Two hierarchical linear feature representations: edge pyramids and edge quadtrees. *Comput. Graph. Image Process.*, **17**, 211.

SMITH, T. R., and PAZNER, M., 1984, Knowledge-based control of search and learning in a large-scale GIS. *Proceedings of the International Symposium on Spatial Data Handling*, Zurich, August 1984, Vol. 2, pp. 498–519.

SMITH, T. R., and PEUQUET, D. J., 1985, Control of spatial search for complex queries in a knowledge-based geographic information system. *Proceedings of the International RSS/CERMA Conference on Advanced Technology for Monitoring and Processing Global Environmental Data*, London, pp. 439–453.

WILKINSON, G. G., and FISHER, P. F., 1984, The impact of expert systems in future operational remote sensing. *Proceedings of the 10th International Conference on Remote Sensing*, Reading, September 1984, pp. 53–60.

Expert Systems for Geographic Information Systems in Resource Management

Vincent B. Robinson,[1] Andrew U. Frank,[2] and Hassan A. Karimi[1]

Abstract

Because environmental information is often spatial, geographic information systems (GIS) are becoming recognized as powerful tools for resource management. GIS are computer-based systems for the collection, storage, management, query, and mapping of geographic data. Problems of GIS to which expert systems have been applied include geographic feature extraction, database systems, decision support, and map design. There remains a large array of areas where use of expert systems would make GIS more directly responsive to the needs of resource managers. These include development of more intelligent user interfaces, better natural language interpretation of field notes, more efficient database search techniques, improved image classification, extraction of high-level geographic features from digitized data, improved management of uncertainty, production of higher quality maps, and use of multiple knowledge bases.

[1] Department of Surveying Engineering, University of Calgary, 2500 University Drive NW, Calgary, Alberta T2N 1N4, Canada (VBRobins@uncamult.bitnet).

[2] Department of Civil Engineering, University of Maine, Orono, Maine 04469 (Frank@mecan1.bitnet).

Much of the information required for resource management is spatial. Conventional information systems are not adept at handling such information or at producing useful maps (Craig 1982). For this reason, geographic information systems (GIS) are becoming recognized as powerful tools for resource management (Webb 1982). A GIS is a computer-based system for collecting, storing, managing, retrieving, transforming, and displaying spatial data from the real world for a particular set of purposes.

The Bureau of Land Management has responsibilities for over 340 million acres of land and resources. As a result of the increasing complexity of their resource management problems, they are committed to using GIS technology to help manage mineral, range, wildlife, and forest resources, to assess the impact of proposed development, and to keep track of land records (Hatch 1986, Parker 1986). Since 1975, the U.S. Fish and Wildlife Service has been developing a GIS for natural resource applications. They now have predictive techniques that model the effects of channel deepening on the distribution of various shellfish and finfish species (Hatch 1986). GIS developed by the National Coastal Ecosystems Team have been used to analyze causes and formulate solutions for wetland change/loss (Ader 1982). The U.S. National Park Service created the

Geographic Information Systems Field Unit, charged with developing, applying, and supporting GIS technology throughout the U.S. National Park system (Fleet 1986). A prototype GIS has recently been evaluated for planning aerial spray block layout for fighting spruce budworm (Jordan and Vietinghoff 1987).

Although GIS technology is being used as a tool for resource management, several areas need improvement. Slow response time results from large volumes of data and the vector overlay approach (Jordan and Vietinghoff 1987). Because most GIS are not particularly easy to use, GIS duties are added to existing responsibilities of already burdened personnel (Fleet 1986). Thus, it is presently unrealistic to expect resource managers to become experts in GIS use. These and other problems of using GIS for resource managment are increasingly being addressed by the application of expert system technology. This paper reviews general aspects of GIS technology, summarizes some research efforts directed towards incorporating expert systems into GIS, and discusses some prospects for the future application of expert systems to GIS.

Geographic Information Systems

Geographic data describe objects from the real world in terms of their 1) position within a known coordinate system, 2) spatial relationships with one another, and 3) attributes (e.g., species composition, soil type). GIS are designed to manage data gathered by widely disparate methods, integrate it, provide analysis, and map the result. A GIS is also supposed to be able to answer a broad range of questions about these data.

Data input encompasses all aspects of transforming data captured in the form of existing maps, field observations, and sensors into a compatible digital form. This process is one of the most costly and time-consuming elements of GIS development. Existing maps, such as those of vegetation distribution, may be captured by a person working with a digitizer to encode the coordinates of desired point, line, and area objects. Scanners have found a niche in automated map input, but they are still not intelligent enough to deal with symbolized lines and some of the problems of text recognition (Dangermond and Morehouse 1987). Aerial photographs still provide much of the information used by resource managers. Capturing environmental data from aerial photographs and encoding them for use in a GIS is a complicated task requiring trained personnel exercising judgments throughout the process. Sensor systems, such as LANDSAT, provide digital image data in a form more readily manipulated by a GIS. Digital map data, like the

U.S. Geological Survey's Digital Line Graph or Digital Elevation Model, are becoming more common as sources of spatial data for GIS.

Geographic data are stored and managed using a variety of structures and methods. Most data storage methods are one of two types--raster (grid cell) and vector (line/polygon). The ability of a GIS to integrate raster data from a satellite sensor with vector data digitized from a map sheet is one of its strengths, as well as the source of many problems in data storage and management. Integration of spatial data from a variety of sources means that problems of locational registration must be solved by the GIS so that all data are located on a common frame of reference. This common spatial frame allows the system to answer queries requiring use of data from a number of different sources. The data integration problem has contributed to making data consistency one of the major problems of GIS (Meier and Ilg 1986).

Query operations in a GIS often involve both the retrieval and analysis of geographic data. Once data are retrieved from the database, the query may require that some kind of spatial analysis be conducted. For example, one may wish to determine the best routing for deploying a forest fire fighting unit. This demands the retrieval of possible routes, as well as some spatial analysis to determine the best route.

Query operations are a series of procedures that make up the question-answering process of a GIS. For example, a simple set of query operations would be to open a data file and then select from that data file those records containing the entry "pine" for the variable tree species. In many of today's GIS, the resource manager would have to specify in detail each of the operations composing a query. Simplifying this interface between GIS and resource manager is where much needs to be done to allow GIS to become more naturally responsive to the manager, rather than requiring the manager to become an expert in GIS operations.

Presentation of the results of geographic query can be accomplished using a range of data output options. However, most users want the graphical output of a GIS to be similar to a conventional map. They require that maps produced conform to all the standards of manual cartography. Thus, even though output techniques of GIS increasingly rely on skills developed in the field of computer graphics, the map remains the most common output product of a GIS. Often users of expensive GIS fail to communicate properly because of a lack of cartographic knowledge on their part or because such knowledge has not been encoded as part of the GIS (Burrough 1986). This deficiency explains why building expert systems for map design is of such great interest to those working to improve GIS.

Expert Systems and Geographic Information Systems

Accomplishments

Expert systems are computer programs that help solve real-world problems normally requiring an expert's interpretation. They are computer models of expert human reasoning. What areas of GIS operation could benefit from the advice of an "expert" to solve real-world problems? Some of these areas can be identified by considering some efforts related to expert systems that have been developed in areas of particular importance to GIS (Table 1). These efforts can be divided into four broad categories: geographic feature extraction, geographic database systems, geographic decision support, and map design.

Table 1. Some expert systems for geographic information systems.

Problem Domain	Expert System	Developers
Map design	MAPEX	Nickerson and Freeman (1986)
	AUTONAP	Freeman and Ahn (1984)
	MAP-AID	Robinson and Jackson (1985)
	ACES	Pfefferkorn et al. (1986)
	CES	Muller et al. (1986)
Geographic feature extraction	ACRONYM	Brooks (1983)
	FES	Goldberg et al. (1984)
	CERBERUS	Engle (1985)
	MAPS	McKeown (1984)
	SPAM	McKeown et al. (1985)
Geographic database systems	ORBI	Pereira et al (1982)
	LOBSTER	Frank (1984)
	KBGIS	Glick et al. (1985)
	KBGIS-II	Smith et al. (in press)
	SRAS	Robinson and Wong (1987)
Geographic decision support	TS-PROLOG	Barath and Futo (1984)
	URBYS	Tanic (1986)
	GEODEX	Chandra and Goran (1986)
	ASPENEX	Morse (1987)
	AVL 2000	Karimi et al. (1987)

Geographic feature extraction. ACRONYM is an expert system that was developed to automate aerial image interpretation (Brooks 1983). ACRONYM incorporates three expert systems, each of which has its own knowledge representation and type of reasoning. All three cooperate to interpret images. The expert systems are: 1) a prediction system using three-dimensional models to predict geometrically invariant features to look for in an image; 2) a description system using an image to obtain descriptions of possible image features; and 3) an interpretation system that uses the descriptions from expert system #2 to find constraints and to check for consistency in the results. Graph matching is used to perform reasoning. The three systems are iterated (prediction to description to interpretation) to get increasingly more detailed interpretations of the image. ACRONYM has been tested on a limited amount of imagery for a small set of objects. Thus, the control mechanisms in ACRONYM are unlikely to be able to handle interpretation of images containing a large set of complex objects (Lambird et al. 1984).

FES (Forestry Expert System) (Goldberg et al. 1984) is used to analyze multi-temporal LANDSAT data for for the detection and monitoring of change in forests. It has been trained to analyze forest cover changes in a 50 by 50 km forested region of Newfoundland, Canada. Using a multi-temporal LANDSAT image database, production rules are applied in two phases. First, rules are used to detect change and estimate reliability. They decide whether the changes in classification from one time period to another are real or if they are statistical artifacts. The second phase generates a set of decision rules regarding the current state of the image. For example, a new decision in the second phase could be arrived at in the following manner: "Given that the previous decision was softwood, and a statistically significant change has occurred, specifically softwood to clearcut, it can be concluded that the area represented has undergone logging."

The control structure of FES is a "feed forward" system. Once a decision has been reached in the second phase, the system starts another iteration. There is apparently no backtracking, but there is a provision for adding production rules by an expert should an unusual situation occur.

FES was tested using imagery covering 100 sq km in central Newfoundland, Canada, over a period of six years. Among the test results was the detection of forest regeneration inferred by changes to areas previously classified as clearcut. Spruce budworm damage was inferred from changes in winter imagery that indicated areas previously classified as softwood were appearing as hardwood. They also report an instance where weather phenomena caused a water area to be misclassified as a burn area, and FES, using its accumulated knowledge, rejected this classification and retained the correct, original classification.

Palmer (1984) uses logic programming (PROLOG) as the basis of an expert system for analysis of terrain features. Using a triangular tesselation of elevation data, Palmer represented nodes with their elevation, and segments and triangles as first-order predicates. Then, using PROLOG to conduct symbolic analyses, he demonstrated how valleys, streams, and ridges could be detected using the procedural knowledge encoded in a knowledge base and using PROLOG control mechanisms. This work was elaborated by Frank et al. (1986) who attempted to show how expert system formalism can be used to define uncertain notions about physical geography. They applied first-order predicate calculus in PROLOG syntax to define geomorphology terms.

Mount Indefatigable in Kananaskis Provincial Park, Canada, with the definition of mountain peak and stream junction represented in PROLOG. The rules displayed above Mount Indefatigable are characteristic of those used in some natural resource expert systems to infer natural phenomena from elevation and location.

CERBERUS, a forward-chaining production rule system incorporating MYCIN-like confidence factors, was developed at NASA to perform unsupervised classification of multispectral data gathered by the LANDSAT satellite (Engle 1985). In the early stages of its development, multispectral data had to be entered by the user, but this is no longer necessary. The FOR-

TRAN-based system is currently being sold as a knowledge-based system for experimenting with expert systems (Digital Review 1986).

Geographic database systems. ORBI, developed to keep track of the environmental resources of Portugal, is another expert system implemented in PROLOG. ORBI includes features of both a classification system for environmental data and a decision support system for resource planning. ORBI provides graphic input and output of maps via a digitizing table and plotter, a natural language parser for Portugese that supports pronouns, ellipses, and other transformations, a menu handler for fixed-format input, an explanation facility that keeps track of the steps in a deduction and shows them on request, and help facilities that explain what is in the database, the kinds of deductions that are possible, and the kinds of vocabulary and syntax that may be used.

LOBSTER (Frank 1984) is a query language for a geographic database (Frank 1982). LOBSTER serves as an intelligent user interface to an object-oriented network spatial database management system. Using the PROLOG syntax to build a user interface to the PANDA database management system (Frank 1984) is an alternative to the extension of standard query languages to include facilities to access GIS. In addition to the intelligent user interface, LOBSTER includes a facility for a programmer to define new "built-in" predicates that are defined as Pascal programs on the object in the database.

SRAS (Robinson and Wong 1987) is a spatial relations acquisition station designed to acquire representations of spatial relations expressed as natural language concepts (such as near, far, close to, remote from) for use in subsequent queries of a geographic database. It addresses the problem of how to represent natural language concepts so that they can be used to answer geographic queries. For example, a resource manager may want to know which streams are "near" a proposed waste disposal site. SRAS would interact with the user to arrive at an explicit representation of the user's meaning of "near."

SRAS is based on a process for acquiring linguistic concepts through mixed-initiative human-machine interactions. The process, using fuzzy topological sets, is designed to determine the meaning of a spatial relation such as "near." First, the system is initialized at a maximum uncertainty condition. Second, SRAS chooses a question so that a measure of expected certainty is minimized. Third, the user's response to the question is used to adjust representation of the linguistic variable. Finally, before starting question selection and concept processing again, a process controller determines whether the fuzziness of the concept has been reduced to an acceptable degree of approximation. This process is unique because it uses fuzzy representations while only requiring the user to answer the questions with a simple yes or no.

SRAS can also be used to acquire multiperson concepts. This ability has particular relevance to the usual organizational context of GIS, where terms and relations are the the product of committee definition or review (Robinson and Wong 1987). This station has potential value for others seeking to develop systems for acquiring spatial knowledge for use in expert systems.

KBGIS is a comprehensive knowledge-based GIS using hybrid knowledge representation (Glick et al. 1985). Frame-based semantic nets are used to represent the geographical objects and their interrelationships. This representation provides the ability to incorporate new entities, attributes, and relationships. These new entities can also inherit characteristics from their object-types or similar entities. Incorporation of geographical entities is facilitated by the use of an expert system shell that is part of KBGIS. KBGIS has recently become operational and is used on a daily basis for trafficability studies and geopolitical trend analysis.

KBGIS-II (Smith and Pazner 1984) is a prototype knowledge-based GIS designed to speed the search through very large geographic data bases. It uses a spatial object language similar to others based on predicate calculus. Using the spatial object language, a user can edit the knowledge base or perform a query operation. The search procedures are concerned with the satisfaction of spatial constraints, primarily whether an object lies within a particular window (e.g., rectangular area). In the database, spatial objects are represented in a specialized hierarchical grid structure called a quad-tree. KBGIS-II also uses some primitive learning procedures to make spatial searches more efficient. Hardware deficiencies do not permit this system to be truly interactive (Smith et al., in press).

Geographic decision support. ASPENEX (Morse 1987) is an expert system to aid management of aspen resources in the Nicolet National Forest, Wisconsin. ASPENEX interfaces an expert system with a GIS. The GIS component is MOSS (Map Overlay and Statistical System), a public domain GIS. The user interface, rule base, and interprogram communication were developed using the shell, EXSYS, on a microcomputer. The microcomputer resident expert system provides rules required in aspen management and MOSS provides spatial information on characteristics of aspen stands.

Barath and Futo (1984) developed a logic-based expert system for comparing requirements of economic sectors (e.g., agriculture, industry, construction) and social factors (e.g., living conditions, employment, education, culture) for making planning decisions. The goals of a district are evaluated in light of the planning actions to be taken, such as building a school, and the constraints of time and cost. Alternative actions may also be evaluated. For example, the system evaluates the feasibility of building a road and a school or just a school.

The system is based on a hierarchical model of national, county, and district-level activities. This goal-directed system was first developed in T-PROLOG and is now based on the TS-PROLOG language. TS-PROLOG is a version of standard PROLOG (Clocksin and Mellish 1981) that has been extended to allow for parallel processes and system simulation. Processes behavior is described by Horn clauses. Therefore, extensions to PROLOG are kept consistent with the representational calculus of logic programming. This system is still an experimental application, primarily funded by the Ministry of Industry of Hungary.

URBYS (Tanic 1986) is an expert system to aid in territorial planning and analysis of urban areas of Martinique. The organization of URBYS is characteristic of other hybrid systems (e.g., Chandra and Goran 1986, Glick et al. 1985). URBYS is organized into a fact base that contains all facts that the interpreter has inferred or that have been provided by the user, a knowledge base that contains rules, the interface to the relational database, and a rule interpreter that serves as an inference engine. The user can forward and/or backward chain. The domain expert changes the rules and/or facts, and URBYS takes no initiative in the knowledge acquisition process.

GEODEX (Chandra and Goran 1986) was built to help planners evaluate site suitability for particular land-use activities. Its rules were drawn from an expert land-use planner. Using the rules in its knowledge base, GEODEX uses a constraint-directed search strategy. For example, when locating a shopping center, constraints such as nearness to a primary road and distance from schools are used to determined the suitability of a site. The land-use planner can develop the complete bundle of constraints, in the form of rules, that apply to the location of a particular land use.

Although the Map Analysis Package (MAP) (Tomlin 1983), a well-known FORTRAN package for geographic information processing of raster data, was used to perform map overlays, GEODEX (written primarily in

LISP) does not interface with any geographic information system. It represents maps in an object-oriented format with each object having a frame of associated information. The frame information is extracted and used by the rule-based system. The map overlays performed by MAP are used to provide information for the frames.

Map design. The MAP-AID expert system being developed in the United Kingdom (Robinson and Jackson 1985) is an attempt to develop a comprehensive expert cartographic system. The goal is to improve the quality of maps derived from GIS by taking complete control of map design and production. MAP-AID is divided into four components: 1) the "core," containing the map design rule base and other information held as rules in the knowledge base, 2) the user module through which the user controls the system and interacts with the knowledge base, 3) a set of data-system modules, and 4) a set of graphics package modules. PROLOG is being used as the primary language for development of this expert system, but the project has faltered due to unanticipated difficulties in formalizing cartographic knowledge.

AUTONAP (Ahn 1984, Freeman and Ahn 1984) is one of the most successful name placement expert systems developed so far. This system emulates an expert cartographer in the task of placing feature names on a geographic map. It uses heuristic knowledge about map placement based on established procedures and conventions. The knowledge base consists of a small set of about 30 explicit rules organized as subroutines in a large RATFOR program on a Prime 750. In AUTONAP, area features are annotated first, then point features, and finally line features. In this manner the system progresses from the most constrained to the least constrained feature annotation task. Some backtracking is part of the system. For example, in point feature annotation, if it becomes impossible to place a name, the system backtracks, removing names already placed and placing them in different positions. Once the placement of area feature names has been accomplished, a free-space list and possible-positions list are developed and used in subsequent name placement tasks. These two lists form a graph of permissible name placements and locations. In point feature name placement tasks, a heuristic graph-searching algorithm similar to the A* algorithm (Nilsson 1971) is used to search the state space. Then line feature names are placed according to a set of rules and constrained by the location of point and area feature names.

MAPEX is a rule-based system that emulates a cartographer in the task of defining and maintaining the high-level decisions of map generalization. Map generalization is the task of generating a map of reduced scale from an existing map (e.g., producing a map of 1:100,000 scale from one of 1:24,000 scale). Map generalization involves deleting features considered insignificant at the reduced scale, simplifying features, combining features, and converting some features from one type to another (e.g., a river with some width converted to one represented by a single line). The generalization process should be accomplished in such a manner that the map effectively communicates the desired information to the user.

MAPEX works with 1:24,000 scale U.S. Geological Survey Digital Line Graph data and is capable of line feature generalization, feature combination, feature deletion, feature simplification, interference detection, feature displacement, and displacement propagation. English-like rules are compiled into a FORTRAN-77 subroutine. To facilitate generation of a formal grammar, generalization actions are described according to data types such as hydrography, transportation, or cultural features. The data types used in the formal grammar of rule building are dictated by conventions used by the U.S. Geological Survey's Digital Line Graph (Allader and Elassal 1983). MAPEX splits automated map generalization tasks into high-level rule definition and translation and low-level geometric manipulation of map data.

CES is a prototype cartographic expert system intended for use by Energy Mines and Resources as an advisor to help cartographers design the Electronic Atlas of Canada. Using decison tables, CES provides a set of mapping specifications that optimally fulfill a set of map requirements given to the system. Preliminary responses from users indicate that cartographers disagree 1) on taxonomy of cartographic concepts that should be introduced in the system, 2) on the rules associating map requirements to map specifications, and 3) on the formal mapping requirements derived from their specific needs. This project experienced some of the same problems as MAP-AID when its developers discovered that cartographic knowledge is difficult to formalize and sometimes inconsistent (Muller 1986, Muller et al. 1986).

Prospects

Some of the most promising applications of expert systems for GIS lie in development of intelligent user interfaces, efficient spatial database search techniques, improved image classification, and production of high quality cartographic products (Ripple and Ulshoefer, in press). There are many other areas suitable for expert system development within each of the major components of GIS--input, management, query, and output--where decisions must be made that will affect both the efficiency and effectiveness of the GIS.

Geographic data input. Geographic data input encompasses all aspects of transforming data captured in the form of existing maps, field observations, and sensors into a compatible digital form. Scanners are being used to transform manuscript maps into digital representations. Contrary to earlier predictions, scanners have not taken over the capture of cartographic data. Dangermond and Morehouse (1987) suggest that the main reason scanners have not taken over is that a lot of map revision is going on at the same time as manuscripts are being captured. For expert systems to automatically capture manuscript maps, they must have knowledge of the scanning system, map object modeling, image analysis, and the map revision process.

Field observations range from the precise measurements of the surveyor to the field notes of a wildlife biologist. There is a need for expert systems to translate these observations into a form useful within a GIS. Digital data recorders are now available for use in surveying operations, but these raw data must be organized and subsequently checked for consistency. The input and transformation of field notes by an expert system would demand a sophisticated understanding of the problem domain language. Of substantial importance would be the ability of the system to understand spatial descriptions at a level permitting the inference of a location in coordinate space.

Lambird et al. (1984) argue that past techniques for the automated or semi-automated extraction of geographic features from remotely sensed images did not work well because of 1) the large amount of information in each image, 2) ambiguous and contradictory information present in the images, 3) problems of dealing with perspective changes, object scale, and resolution, 4) many sources of geometric and radiometric variability, and 5) lack of adequate models relating physical principles to object appearance in images. The advent of expert systems holds hope that many of these problems may be solved. Expert systems that interpret images require knowledge about image processing, image formation, object recognition, and object modeling. This knowledge is generally distributed among several experts. Expert systems like ACRONYM may exploit the advantages of distributed problem solving.

There is an increasing amount of data that is not of image form but is best characterized as digital-spatial data. The spatial data provided in digital form may not contain the features a resource manager finds useful, but may contain the data from which those features must be inferred. For example, much resource management is oriented around stream networks and watersheds. Band (1986) has shown one way that stream networks and their accompanying watersheds can be extracted from a digital elevation model. However, the extraction of high-level geographic features requires not only knowledge about their geometry, but also domain specific expertise such as that possessed by physical geographers, ecologists, and foresters.

Geographic data storage and management. Although systems such as URBYS are exploiting existing data bases, it is clear from the systems discussed above that the methods of storing geographic data are changing in response to the demands of expert systems. In an effort to develop storage structures that support rapid spatial searches, some have begun to base their systems upon the quad-tree (Antony and Emmerman 1986; Smith et al., in press). Regardless of the specific storage model, several areas of geographic data management require significant expertise.

The ability of a GIS to integrate a number of data layers is one of its strengths. However, this is also the source of many problems. Expert systems developed to manage the geographic database will need knowledge of positional modeling, spatial interpolation, geographic consistency, and uncertainty modeling. The geographic database manager needs to be able to assess the uncertainty characteristics of incoming data and take appropriate action. Many experts in GIS have commented that some people are adept at increasing the certainty of resource data using a reasoning process whereby they apply their expertise and knowledge about ancillary data to arrive at an improved data layer. (An example might be where there is uncertain data about tree species in an area with mountains and valleys. Since some species tend to be located at particular elevations on the windward side of a mountain, use of elevation and location on the mountainside can help improve the certainty with which we can make statements about the distribution of tree species.)

It is, however, important that the various data layers be consistent with one another. Otherwise, logical inconsistencies may arise in subsequent analyses. Meier and Ilg (1986) have discussed some general rules of consistency in a simple spatial database. Much remains to be done to formalize the rules needed for maintaining consistency in a complex, multilayered geographic database.

One of the keys to development of these expert geographic database managers is managing the different kinds of uncertainty found in geographic data bases (Robinson and Frank 1985). Some of developments in this area are seen in the inclusion of a reliability measure in FES, Shine's (1985) review of the utility of Bayesian, fuzzy, and belief logics in feature extraction systems, and Robinson's (in press) discussion of the implications of fuzzy logic for geographic data bases.

Geographic query. Both the retrieval and analysis of geographic data are often part of a geographic query. Retrieval of geographic data demands that some kind of spatial search be conducted. Increasing the efficiency of spatial search is one of the major objectives of KBGIS-II. However, due to reported hardware limitations, KBGIS-II is not truly interactive (Smith et al., in press).

An intelligent user interface could guide an inexperienced user through the most efficient use of the system (Jackson and Mason 1986), but the interface would have to be an expert on that particular GIS. Resource managers, who have little time to become experts in GIS as well as in resource management, would welcome such an interface. An intelligent user interface is one of the concepts behind development of ASPENEX (Morse 1987).

One of the problems in developing an expert system interface is to provide a natural means of communication between the manager and the GIS. This communication demands knowledge of what is meant by the manager's request. Thus, semantic modeling of resource management queries is needed to develop formal knowledge of how to interpret queries. Robinson and Wong (1987) have shown how seemingly trivial queries can vary, depending upon the user. In controlled, experimental sessions with SRAS, none of five subjects agreed on the meaning of "far from Douglas, Georgia." In fact, a user with graduate-level training in computer cartography gave answers that resulted in an inconsistent meaning of "close to Douglas." In other words, the user said one settlement was close to Douglas, which is farther away than another settlement which as said to be not close to Douglas. Furthermore, out of five subjects, eight terms used to define relations, and two data bases, there was, on the average, less than 50-percent agreement on the meaning of terms describing spatial relations.

Map design. The map remains one of the most common output products of a GIS. However, users of expensive GIS sometimes fail to communicate properly because of a lack of cartographic knowledge (Burrough 1986). MAP-AID and CES both encountered problems due to the informal nature of current cartographic knowledge. The formalization of cartographic knowledge remains one of the challenges of constructing carto-

graphic expert systems (Fisher and Mackaness 1987). As discussed above, progress is being made in both automatic name placement and map generalization

Another aspect of the map design process is identification and resolution of spatial conflicts when placing symbols on a map. Mackaness and Fisher (1987) have attempted to formalize the process of automatic recognition and resolution of spatial conflict in cartographic symbolizations, but further research is required to resolve some of the problems of overlap, enclosure, and worst case resolution that were identified by their investigations. Like others working on cartographic expert systems, their goal is to enable researchers with no cartographic skills to display field data using designs that best communicate those data.

Conclusions

The self-knowledge or explanatory capability of cartographic expert systems is presently critical to their successful use because automated methods of deciding among competing cartographic designs have yet to be developed (Fisher and Mackaness 1987). Formalizing knowledge of this type needs to receive substantial emphasis in future development of expert GIS.

Development of smaller prototypes using structured knowledge acquisition methodologies and semantics will not only bring practical results, but, more importantly, will help clarify many less formalized areas. Lessons learned in building these smaller systems will in turn be transportable and expandable to later more advanced systems. By developing more than one prototype addressing the same problem but using different methods, we can begin to conduct comparative analyses that can support more informed decisions concerning the development of expert systems for GIS.

Acknowledgments

This research received partial support from the Alberta Forestry, Lands, and Wildlife Professorship in Digital Mapping and Spatial Data Management, National Science Foundation grants IST-8412406 and IST-8609123, and a grant from the Natural Sciences and Engineering Research Council of Canada.

References

Ader, R. R. 1982. A geographic information system for addressing issues in the coastal zone. Computers, Environment, and Urban Systems 7(4): 233-

244.

Ahn, J. K. 1984. Automatic map name placement. IPL-TR-063, Image Processing Laboratory, Rensselaer Polytechnic Institute, Troy, New York.

Allader, W.R., and A. A. Elassal. 1983. Digital line graphs from 1:24000-scale maps. United States Geological Survey Circular 895-C, United States Geological Survey, Reston, Virginia.

Antony, R., and P. J. Emmerman. 1986. Spatial reasoning and knowledge representation. Pages 795-814 in: Geographic information systems in government, B. Opitz, editor. A. Deepak Publishing, Hampton, Virigina.

Band, L. E. 1986. Topographic partition of watersheds using digital elevation models. Water Resources Research 22: 15-24.

Barath, E., and I. Futo. 1984. A regional planning system based on artificial intelligence concepts. Papers and Proceedings of the Regional Science Association 55: 135-154.

Brooks, R. A. 1983. Model-based three-dimensional interpretations of two-dimensional images. IEEE Transactions on Pattern Analysis and Machine Intelligence, PAMI-5: 140-150.

Burrough, P. A. 1986. Principles of geographic information systems for land resources assessment. Oxford University Press, New York.

Chandra, N., and W. Goran. 1986. Steps toward a knowledge-based geographical data analysis system. Pages 749-764 in: Geographic information systems in government, B. Opitz, editor. A. Deepak Publishing, Hampton, Virginia.

Clocksin, W., and C. Mellish. 1981. Programming in PROLOG. Springer-Verlag, New York.

Craig, W. J. 1982. Preface. Computers, Environment, and Urban Systems 7(4): 213-218.

Dangermond, J., and S. Morehouse. 1987. Trends in hardware for geographic information systems. Pages 380-385 in: Proceedings, Eighth International Symposium on Automated Cartography, Baltimore, Maryland, March 1987.

Digital Review. 1986. New software product report on CERBERUS, p. 188.

Engle, S. W. 1985. CERBERUS release notes version 1.0, NASA Ames Research Center, Moffett Field, California.

Fisher, P., and W. Mackaness. 1987. Are cartographic expert systems possible? Pages 530-534 in: Proceedings, Eighth International Symposium on Automated Cartography, Baltimore, Maryland, March 1987.

Fleet, H. 1986. Geographic information systems and remote sensing activities in the National Park Service. Pages 635-644 in: Geographic information systems in government, B. Opitz, editor. A. Deepak Publishing, Hampton, Virigina.

Frank, A. U. 1982. MAPQUERY: data base query language for retrieval of geometric data and their graphical representation. Computer Graphics 16: 199-207.

Frank, A. U. 1984. Extending a network database with PROLOG. Pages 665-674 in: Proceedings, First International Workshop on Expert Database Systems, Kiawah Island, South Carolina, October 1984.

Frank, A. U., B. Palmer, and V. B. Robinson. 1986. Formal methods for accurate definitions of some fundamental terms in physical geography. Pages 583-599 in: Proceedings, Second International Symposium on Spatial Data Handling, Seattle, Washington, July 1986.

Freeman, H., and J. Ahn. 1984. AUTONAP--an expert system for automatic map name placement. Pages 544-571 in: Proceedings, First International Symposium on Spatial Data Handling, Zurich, Switzerland, August 1984.

Glick, B., S. A. Hirsch, and N. A. Mandico. 1985. Hybrid knowledge representation for a geographic information system. Paper presented at the Seventh International Symposium on Automated Cartography, Ottawa, Ontario, March 1985.

Goldberg, M., M. Alvo, and G. Karam. 1984. The analysis of LANDSAT imagery using an expert system: forestry applications. Pages 493-503 in : Proceedings, Sixth International Symposium on Automated Cartography, Ottawa, Ontario, March 1984.

Hatch, H. J. 1986. Overview of geographic information systems in the federal government. Pages xvii-xx in: Geographic information systems in government, B. Opitz, editor. A. Deepak Publishing, Hampton, Virginia.

Jackson, M. J., and D. C. Mason. 1986. The development of integrated geo-information systems. International Journal of Remote Sensing 7: 723-740.

Jordan, G., and L. Vietinghoff. 1987. Fighting spruce budworm with a GIS. Pages 492-499 in: Proceedings, Eighth International Symposium on Automated Cartography, Baltimore, Maryland, March 1987.

Karimi, H. A., E. J. Krakiwsky, C. Harris, G. Craig, and R. Goss. 1987. A relational database model for an AVL system and an expert system for optimal route selection. Pages 584-593 in: Proceedings, Eighth International Symposium on Automated Cartography, Baltimore, Maryland, March 1987.

Lambird, B. A., D. Lavine, and L. N. Kanal. 1984. Distributed architecture and parallel non-directional search for knowledge-based cartographic feature extraction systems. International Journal of Man-Machine Studies 20: 107-120.

Mackaness, W. A., and P. F. Fisher. 1987. Automatic recognition and resolution of spatial conflicts in cartographic symbolization. Pages 709-718 in: Proceedings, Eighth International Symposium on Automated Cartography, Baltimore, Maryland, March 1987.

McKeown, D. M. 1984. Digital cartography and photointerpretation from a database viewpoint. Pages 19-42 in: New Applications of Databases, G. Gargarin and E. Colembe, editors. Academic Press, New York.

McKeown, D. M., W. A. Harvey, and J. McDermott. 1985. Rule-based interpretation of aerial imagery. IEEE Transactions on Pattern Analysis and Machine Intelligence PAMI-7: 570-585.

Meier, A., and H. Ilg. 1986. Consistent operations on a spatial data structure. IEEE Transactions on Pattern Analysis and Machine Intelligence PAMI-8: 532-542.

Morse, B. 1987. Expert interface to a geographic information system. Pages 535-541 in: Proceedings, Eighth International Symposium on Automated Cartography, Baltimore, Maryland, March 1987.

Muller, J.-C. 1986. Personal communication. University of Alberta, Edmonton, Alberta, November 1986.

Muller, J.-C., R. D. Johnson, and L. R. Vanzella. 1986. A knowledge-based approach for developing cartographic expertise. Pages 557-571 in: Proceedings, Second International Symposium on Spatial Data Handling, Seattle, Washington, July 1986.

Nickerson, B. G., and H. Freeman. 1986. Development of a rule-based system for automatic map generalization. Pages 537-556 in: Proceedings, Second International Symposium on Spatial Data Handling, Seattle, Washington, July 1986.

Nilsson, N. J. 1971. Problem-solving methods in artificial intelligence. McGraw-Hill, New York.

Palmer, B. 1984. Symbolic feature analysis and expert systems. Pages 465-478 in: Proceedings, First International Symposium on Spatial Data Handling, Zurich, Switzerland, August 1984.

Parker, H. D. 1986. Geographic information systems technology in natural resource management: the Bureau of Land Management's example. Pages 1-6 in: Geographic information systems in government, B. Opitz, editor. A. Deepak Publishing, Hampton, Virginia.

Pereira, L. M., P. Sabatier, and E. de Oliveira. 1982. ORBI--An Expert System for Environmental Resource Evaluation through Natural Language. Report FCT/DI-3/82, Departmento de Informatica, Universidade Nove de Lisboa.

Pfefferkorn, C., D. Burr, D. Harrison, B. Heckman, C. Oresky, and J. Rothermel. 1985. ACES: a cartographic expert system. Pages 399-407 in: Proceedings, Seventh International Symposium on Automated Cartography, Washington, D.C., March 1985.

Ripple, W. J., and V. S. Ulshoefer. In press. Expert systems and spatial data models for efficient geographic data handling. Photogrammetric Engineering and Remote Sensing 53(10).

Robinson, V. B. In press. Some implications of fuzzy set theory applied to geographic databases. Computers, Environment, and Urban Systems.

Robinson, G., and M. Jackson. 1985. Expert systems in map design. Pages 430-439 in: Proceedings, Seventh International Symposium on Automated Cartography, Washington, D.C., March 1985.

Robinson, V. B., and A. U. Frank. 1985. About different kinds of uncertainty in geographic information systems. Pages 440-450 in: Proceedings, Seventh International Symposium on Automated Cartography, Washington, D.C., March 1985.

Robinson, V. B., and R. N. Wong. 1987. Acquiring approximate representations of some spatial relations. Pages 604-622 in: Proceedings, Eighth International Symposium on Automated Cartography, Baltimore, Maryland, March 1987.

Shine, J. A. 1985. Bayesian, fuzzy, belief: which logic works best? Pages 676-679 in: Proceedings, American Society of Photogrammetry, Washington, D.C., March 1985.

Smith, T. R., D. Peuquet, and S. Menon. In press. KBGIS-II: a knowledge-based geographic information system. International Journal of Geographic Information Systems.

Smith, T. R., and M. Pazner. 1984. Knowledge-based control of search and learning in a large-scale GIS. Pages 498-529 in: Proceedings, First International Symposium on Spatial Data Han-

Robinson et al.: *Geographic Information Systems*

dling, Zurich, Switzerland, August 1984.

Tanic, E. 1986. Urban planning and artificial intelligence; the URBYS system. Computers, Environment, and Urban Systems 10(3-4): 135-146.

Tomlin, D. 1983. Digital cartographic modelling techniques in environmental planning. Ph.D. dissertation, Yale University, New Haven, Connecticut.

Webb, P. R. 1982. A synopsis of natural resource management and environmental assessment techniques using geographic information system technology. Computers, Environment, and Urban Systems 7(4): 219-232.

Andrew U. Frank holds a Dipl. Ing. and a doctorate of technical sciences from the Swiss Institute of Technology, Zurich, Switzerland (ETH). His thesis, Data Structures for Land Information Systems--Semantical, Topological, and Spatial Relations in Geo-scientific Data, concentrated on the use of database management systems for spatial data. Frank is an Associate Professor at the University of Maine, where he has taught courses in spatial information systems since 1983. He designed and implemented PANDA, a spatial database system. Currently he is investigating methods of building more appropriate models of reality in databases using methods and results from artifical intelligence research. His main research interest is in the representation of geometric knowledge in information systems.

Vincent B. Robinson received his Ph. D. in geography in 1978 from Kent State University. Robinson has traveled widely as a consultant on spatial data management and systems analysis for international organizations. He developed the Urban Data Management Software (UDMS) package for the United Nations Centre for Human Settlements and has conducted research projects for NASA and NSF. Since 1987, Robinson has been Associate Professor in the Department of Surveying Engineering, University of Calgary, and holds the Alberta Forestry, Lands, and Wildlife Professorship in Digital Mapping and Spatial Data Management. He teaches courses on principles of intelligent land data systems, artificial intelligence in surveying and mapping, and land information systems. His current research areas are expert systems, uncertainty in land information systems, distributed land data systems, and geographic information systems. He recently set up a small project with the Alberta Resource Evaluation Branch of Lands, Forestry, and Wildlife to investigate applying expert systems and GIS for resource evaluation.

Dr. Vincent B. Robinson may now be reached at the following address: Director, Institute for Land Information Management, University of Toronto, Erindale College, Mississauga, Ontario L5L 1C6, Canada (Bitnet: vbr__ilim@erin.utoronto.ca).

Hassan A. Karimi holds a B. Sc. from the University of New Brunswick and an M.Sc. from the University of Calgary in computer science. His thesis, A Preprocessor Approach for Distributed Databases in INGRES, involved the design and development of a preprocessor for distributed relational databases. He has taught courses in computer science and database managment. Karimi is currently pursuing his Ph.D. in survey engineering and digital mapping. As a research assistant, he has been instrumental in the development of both land information systems (LIS) and automatic vehicle location (AVL) systems. His current research areas are the formalization and development of intelligent systems such as AVL, expert systems in the domain of land data systems, and distributed spatial information systems.

Current and potential uses of geographical information systems

The North American experience

TOMLINSON ASSOCIATES, LTD.

Tomlinson Associates Ltd, 17 Kippewa Drive, Ottawa, Ontario, Canada

Abstract. This paper provides an overview of developments and applications of geographical information systems (GIS) in North America over the past 20 years, together with some indications of lessons learned and prospects for the future. Most developments in this field have been confined to a small number of sectors, including forestry, land registry, transport and facility planning, civil engineering, agriculture and environment, although progress has been variable. The wider use of GIS has been (and will probably continue to be) inhibited by a number of factors, such as lack of digital base mapping, limited digital data, unawareness among potential users and resistance to new technology, coverage of data, assessment of costs and benefits, technological developments affecting GIS and sources of funds for research to political structures, availability of skilled staff, lack of proper advice, the unwisdom of 'going it alone' and management problems (which are as important as technical problems, if not more so). A successful programme of GIS depends on a coordination of effort relating to applications, trained personnel, governmental involvement in R. & D. and development of a source of independent advice.

1. Introduction

Although North America has played the leading role in the development and applications of geographical information systems (GIS), it is difficult to give an overview of the North American experience. Two countries of continental extent, federal systems of government with considerable devolution to states and provinces, a vigorous private sector and, in the United States in particular, a dislike of big government, inevitably imply great variety and complexity. No comprehensive appraisal has been attempted by either federal government and much of the experience is not accessible in print. This overview is thus necessarily subjective, reflecting the experience of Tomlinson Associates Ltd extending over more than 25 years, with strong inputs from academic research and the private sector and from close collaboration with official agencies of one kind and another. None the less, such an attempt seems worthwhile and the Committee of Inquiry into the Handling of Geographic Information (1987) appointed by the British Government commissioned from Tomlinson Associates Ltd the appraisal on which this article is based. In it, the author seeks to provide both an overview of what has happened over the past 20 years, with particular emphasis on the last decade, and some indication of lessons that can be learned and of prospects for the future.

1.1. *Terminology*

Before examining the experiences of the different sectors in which initiatives in developing and using GIS have been taken, it is first necessary to define the terminology used and to clarify the distinction between GIS and related activities, especially automated cartography; for a lack of understanding of these differences has been the source of much difficulty in the past. Geographical information systems are concerned

© 1987, Taylor & Francis, Ltd., *International Journal of Geographical Information Systems*, v. 1, n. 3, p. 203–218; reprinted by permission.

with the handling of geographical data, which are those spatial data that result from observation or measurement of earth phenomena (spatial data being those that can be individually or collectively referenced to location). A GIS is a digital system for the analysis and manipulation of a full range of geographical data, with associated systems for inputting such data and for displaying the output of any analyses and manipulations. In geographical information systems the emphasis is clearly on these latter functions, which provide the main motivation for using digital methods. Automated cartography, on the other hand, is the use of computer-based systems for the more efficient production of maps, and while maps may often be the form selected for output from a GIS, the data structures for GIS and for automated cartography and the functions they provide are different and the two types of systems are not highly compatible.

2. Applications of GIS by sector

GIS have been developed independently for a wide variety of purposes and the future of GIS will depend to a large extent on the degree to which these various needs can be integrated and met by one type of product. The growth of GIS in recent years has been led by developments in a small number of sectors and there have been distinct differences in the forms that development has taken and in the meanings attached to GIS. The sectors which are discussed below represent those which Tomlinson Associates Ltd believe to have played an important role in the development and use of GIS technology in North America.

2.1. Forestry

Forestry has been responsible for a significant growth in the use of GIS in the past 5 years. Ideally, GIS technology would be used for updating and maintaining a current forest inventory and for modelling and planning forest management activities such as cutting and silviculture, road construction and watershed conservation; in other words, the true advantages of GIS accrue only when emphasis is placed on manipulation, analysis and modelling of spatial data in an information system. In practice GIS have often been used for little more than automation of the cartography of forest inventories, because of limitations in the functionality of software or resistance to GIS approaches on the part of forest managers. This situation will change slowly and reflects a general problem also affecting sectors other than forestry.

In a typical North American forest management agency the primary cartographic tool for management is the forest inventory. It is prepared for each map sheet in the agency's territory on a regular cycle, which requires flying and interpreting aerial photography, conducting operational traverses on the ground and manual cartography. In one such agency the number of sheets is 5000 and the cycle of updating is 20 years. Events which affect the inventory, such as fire, cutting and silviculture, are not added to the basic inventory.

The earliest motivation for establishing GIS in forestry was the ability to update the inventory on a continuous basis by topological overlay of records, reducing the average age of the inventory from the existing 10 years to a few weeks. More sophisticated uses include calculation of cuttable timber, modelling outbreaks of fire and supporting the planning of management decisions. Every significant forest management agency in North America either has now installed a GIS, or is in some stage of acquiring one. No agency is known to have rejected GIS in the past 3 years.

The number of installed systems in this sector in 1986 is estimated at 100, in federal, state and provincial regulatory and management agencies and in the private sector of the forest industry. This figure is based on a census of Canadian agencies, where 13 are known to exist at present, and an estimate of the relative sizes of the Canadian and U.S. industries. Systems have been supplied by the private sector (e.g. Environmental Systems Research Institute (ESRI), Comarc, Intergraph) or developed internally (e.g. Map Overlay Statistical System (MOSS)), and there are also systems in the public domain with various levels of support from the private sector (MOSS, several raster-based systems). Many of these systems do not support a complete range of forms of spatial analysis on points, lines, grid cells, rasters and irregular polygons. Almost all major agencies and companies in the North American forest products industry would now claim some level of involvement in GIS: the figure of 100 is an estimate of those that currently maintain a significant establishment of hardware, software and personnel with recognizable GIS functions.

Several forms of *de facto* standardization have emerged in various parts of the industry. In New Brunswick the private and public sectors have coordinated efforts by acquiring identical systems: the New Brunswick Department of Natural Resources, Forest Management Branch, and the two main forest products companies have all acquired ESRI systems. The U.S. Forest Service is in the process of determining a coordinated approach to GIS for its ten regions and 155 forests.

The potential market in North America is approximately 500, on the assumption that all U.S. National Forests will acquire a minimal system, together with the regional and national offices of the Forest Service, and that in large agencies such as the Forest Resources Inventory of the Ontario Ministry of Natural Resources there will be systems in each of the major regions. In the absence of major downturns in the industry this level is expected to be reached within the next 5 to 10 years. At the same time the emphasis will shift increasingly to analytic rather than cartographic capabilities, so that there will be a growing replacement of simple systems by those with better analytic functionality. This will be an expensive change for some agencies which will find that their early systems or data bases cannot be upgraded, often because of the lack of topology in the data structures.

Several general points can be drawn from this experience. Firstly, virtually all forest management agencies which have attempted to 'go it alone' and develop systems, either in-house or through contract, have met with disaster, in some cases repeatedly. The development of a GIS is a highly centralized function requiring resources far beyond those of a single forest management agency, an observation that has a much wider relevance. Secondly, the resources of a single agency do not permit effective, informed evaluation of commercial products. If acquisition is uncoordinated and agencies are permitted to go their own way, the result is a diversity of incompatible systems which are generally inappropriate solutions to their needs for GIS. There is very great variation both in the capabilities of commercial systems and in the needs of each agency, and matching these is a difficult and complex task. As the field matures the differences between systems will presumably lessen, but the variation in the nature of each agency's workload will, if anything, increase in response to more and more sophisticated forms of analysis and modelling. There is thus no such thing as a common denominator system for the forest industry: a natural tendency is for diversity in the needs of agencies and the responses of vendors, rather than uniformity.

Part of the reason for this diversity lies in the mandates under which many responsible agencies must operate. The typical U.S. National Forest has a mandate to manage not only the forests in its area, but also the wildlife, mineral and recreational resources. Management objectives must thus allow for the need to conserve as well as to extract. The typical Forest Service GIS will be used to manage road facilities, archaeological sites, wildlife habitats and many other geographical features. The relative emphasis on each of these varies greatly from forest to forest; several forests, for example, have significant coal reserves and others have no trees.

There has been no attempt at national level to coordinate GIS technology in the forest industry in either the U.S. or Canada. In Canada the Forestry Service of the Federal Government has an advisory mandate and its acquisition of an ESRI system has possibly had some influence as a centre of excellence on developments in provincial agencies and companies, but this is far from certain. No overall study of sectoral needs is known to exist in either country, nor is there any effort to promote or fund system development to meet those needs.

Several factors account for the recent very rapid growth of GIS activity in the forest industry. Firstly, effective forest management has been a significant societal concern and has attracted government funding. Secondly, GIS technology is seen as an effective solution to the problem of maintaining a current resource inventory, since reports of recent burns, cutting and silviculture can be used to update a digital inventory immediately, resulting in an update cycle of a few months rather than years. Thirdly, a GIS is attractive as a design tool to aid in scheduling cutting and other management activities. Finally, because of the multi-thematic nature of a GIS data base, it is possible to provide simultaneous consideration of a number of issues in developing management plans. All of these factors, combined with the perception that GIS technology is affordable, have given vendors a very active market in the past 5 years or so in North America. It should be noted that, in the last two factors at least, the functions offered by the GIS are substantially new and do not represent automation of an existing manual process.

2.2. *Property and land parcel data*

The acronyms LIS and LRIS (land information system and land related information system respectively) are often used in this sector, reflecting the relative importance of survey data and the emphasis on retrieval rather than on analysis. Most major cities and some counties have some experience in building parcel systems, often dating back to the earliest days of GIS, but state or national systems have not generally been considered in North America because land registration is usually a local responsibility. Fragmentation of urban local government is also a problem in the U.S. and in some Canadian provinces. Nevertheless, the special cadastral problems of the maritime provinces have led to investment there in an inter-provincial system (Land Registration Information System (LRIS), using CARIS software by Universal Systems Ltd).

There is a potential market of some 1000 systems among the 500 major cities with populations of 50 000 or more, the 3000 counties of the U.S. and perhaps a tenth of that number in Canada, although the number of such agencies with significant investment in GIS has not changed markedly in the past 10 years. Although several cities established an early presence in the field, their experiences were often negative and projects were frequently abandoned. More recently, improved software and cheaper hardware have meant a greater rate of success, but this sector is still at a very low point on the growth curve.

The functions needed in a LIS are well short of those in a full GIS. In many cases all that is needed is a geocoding of parcels to allow spatial forms of retrieval: digitizing the outlines of parcels is useful for cartographic applications. It is unlikely that many municipalities will advance to the stage of creating a full urban GIS by integrating data on transport and utilities and developing applications in urban development and planning, at least in the next 5 years. Those where this has been attempted have discovered the difficulties of working across wide ranges of scales, from the 1 : 1000 or greater of data on land parcels to the 1 : 50 000 suitable for such applications as planning emergency facilities, scheduling bus systems and developing shopping centres. There is very little suitable applications software for such activities at present. In summary, automated cartography and retrieval will probably remain the major concerns of such systems in the immediate future and confidentiality and local responsibility will remain barriers to wider integration. It is likely that these needs will be met initially by vendors of automated cartography systems (such as Intergraph and Synercom), and data base management systems (DBMS), although there will be a steady movement towards GIS capabilities as urban planners and managers demand greater analytic capabilities. It is unlikely, however, that many municipalities will reach the stage of giving the digital data legal status as a cadastre because of problems of accuracy and confidentiality.

The influence of government in this sector has largely been driven by concern for the quality of the data, rather than by technical development. Several acquisitions of systems have been sponsored by senior governments for demonstration purposes and as a means of promoting quality and format standards. However, with a few exceptions at a very early stage, there have been no major developments of systems undertaken within this field, most needs apparently being satisfied by hardware and software already available. A major experiment is currently under way in Ontario to apply ESRI software to the needs of municipal data bases for such varied applications as policing, emergency services, transport systems and schools. But it is too early to know whether this mode will lead to a widespread adoption of similar approaches in other cities and to a significant movement up the growth curve.

2.3. *Utilities*

Telephone, electrical and gas utilities operate in both private and public sectors in North America, and it is useful to distinguish between applications at large and small scales. Large-scale applications include monitoring the layouts of pipelines and cables and the location of poles and transformers, and, as in the case of land parcel systems, combining needs for cartography and spatial retrieval. Traditionally these have similarly been met by vendors of automatic cartography systems, particularly Synercom, and by DBMS. However, it is possible that demand for more sophisticated forms of spatial retrieval and layout planning will lead, in the long term, to a reorientation towards the GIS model.

Small-scale applications include the planning of facilities and transmission lines to minimize economic, social and environmental costs, and demand forecasting. Some utilities have built large data bases for such purposes and also make use of digital topographic data. Such applications are highly specialized and idiosyncratic, and have largely been handled by specialized software rather than by generalized systems. In the long term, as GIS software stabilizes and develops, it is likely that the advantages of better capabilities for input and output and easy exchanges of data formats will make such systems an attractive option for these applications. Developments in this sector have not yet reached the beginning of the growth curve, but a significant proportion of this market will probably move away from existing automated cartographic systems into GIS in the next 5 to 10 years, and several vendors appear to have anticipated this trend already, as illustrated by Synercom's marketing of Odyssey. There are some 200 utilities in North America with potential interest in GIS applications.

2.4. *Transport, facility and distribution planning*

In addition to both public and private sector transport agencies, much of the work in this sector is carried out by research contractors, such as market research firms and university staff. Raker (1982) lists 26 companies in this sector providing either GIS services or geographical data, or both. Although market research frequently calls for sophisticated forms of spatial analysis, such as site selection and spatial interaction modelling, vendors of GIS have not yet made any significant penetration of the market. Instead, most companies rely on a combination of standard statistical packages (e.g. Statistical Package for the Social Sciences (SPSS) and Statistical Analysis System (SAS)), data base management systems and thematic mapping packages developed in-house or acquired from vendors (e.g. Geographic Information Manipulation and Mapping System (GIMMS)). Sales of software and data often make a significant contribution to income.

There is every indication that this sector will be a major growth area for applications of GIS in the next 10 years. Software for these forms of spatial analysis is at present rudimentary but developing rapidly (e.g. ESRI's NETWORK). There is a pressing need for the ability to handle multiple formats of geographical data, scales and types of features, and hierarchical aggregation of features, in processing socio-economic data, in combination with advanced forms of spatial analysis and map display. In addition to market research firms, the potential market includes major retailers, school systems, transport and distribution companies and other agencies which need to solve problems of routing and rescheduling on networks, and the direct mail industry. It is not unreasonable to visualize a potential market of 1000 systems in North America, or even ten times that number, but there has been so far very little interaction between vendors of GIS and those software houses which have traditionally operated in this market.

2.5. *Civil engineering*

A major use of digital topographic data is in large-scale civil engineering design, such as cut and fill operations for highway construction. The first digital developments in this field derived from photogrammetric operations, which are the primary source of data. More recently, efforts have been made to add more sophisticated capabilities to photogrammetric systems, notably by Wild (System 9) and Kern, and to interface them with automated cartographic systems and GIS. Government agencies in North America (U.S. Geological Survey (USGS), Canada Department of Energy, Mines and Resources (EMR)) have been fairly active in this field in attempting to develop common data standards and formats for the exchange of data.

There are some 50 major systems installed in civil engineering contractors and government agencies in Canada and perhaps ten times that number in the U.S. This figure is unlikely to change in the next 10 years as this is not a growth industry. However, rapid growth is occurring in both Canada and the U.S. in the significance of digital topographic data for defence, because of its role in a number of new weapon systems, including Cruise, and because of the general increase of defence budgets in the industrialized world. This work has drawn attention to the importance of data quality and to the need for sophisticated capabilities for editing topographic data as well as for acquiring them. These needs are presently being met by enhancements to automatic cartography systems (e.g. Intergraph) and it is not yet clear whether they will lead in the long term to any significant convergence with GIS.

2.6. *Agriculture and environment*

From a Canadian perspective these are the original application areas of GIS: projects in federal agencies with responsibilities for the environment and agriculture, in the form of the Canada Geographic Information System (CGIS) and Canadian Soil Information System (CanSIS) respectively, were started over 20 years ago. The use of GIS approaches can be traced to the need to measure the area of land resources, to reclassify and dissolve prior to display, and to overlay data sets and to compare them spatially. These remain among the most basic justifications for GIS technology. Although both systems were developed largely in-house in government agencies, the capabilities of CanSIS are now provided by a number of cost-effective commercial products. On the other hand, CGIS remains unique: no vendor has yet developed a system with comparable capabilities in bulk data inputting and in archiving national data.

The environmental market is much less significant in most countries than it was 15 years ago, except in specialized fields. GIS technology is of considerable interest in land management, particularly of national parks and other federal, state and provincial lands, and has been adopted in both the U.S. and Canada: there is a potential for perhaps ten such systems in Canada and at least ten times that number in the federal land management agencies in the U.S. However, in the present mood of fiscal restraint in both countries, neither figure is likely to be reached in the immediate future.

In agriculture the main issue arises from the critical importance in farming of changes over time, both seasonal and annual. Although much research has been conducted on the interpretation of agricultural data from remotely-sensed imagery, and there are no major technical problems in interfacing image processing systems with GIS, there remain the conceptual problems of classification and interpretation. Designers of GIS have typically assumed that data are accurate: to date no significant progress has been made in designing systems to process and analyse uncertain data. Similarly, the conventional GIS has no explicit means of handling data which are time-dependent or longitudinal, yet these are characteristic of data in a number of sectors besides agriculture. Marine environmental monitoring is a good example, as is climatology.

3. Inhibiting factors

The wider use of GIS technology in North America has clearly been inhibited by a number of factors and many will continue to have an effect for some time. Firstly, there is no overall programme of large-scale digital base mapping and as yet no suitable base is widely available. This is an inhibiting factor, but not a major one. To be suitable, a digital base would have to exist at a number of scales, since it would be too elaborate and expensive for users to derive specific scales by generalization from a common denominator scale. It would, therefore, have to exist in a number of formats and, since it would be so large that no average user could expect to maintain a private copy, it would have to be available at very short notice, probably on line. Such base maps as exist, notably files of state and county outlines, have been widely distributed and have organizations, journals and textbooks, and there has been no systematic approach to GIS in the educational system. The term is encountered in courses in departments of geography, surveying and forestry, in other words primarily in applications of GIS, by students with little or no technical background. There have been attempts in North America in various sectors to coordinate the development of specialist programmes, but these will not be successful as long as they are perceived as benefiting one sector at the expense of another. The federal governments are unlikely to have much influence on the educational system until they can coordinate their own interest in geographical data handling, and there is little sign of their doing so at present. The greatest progress has been made by provincial governments in Canada, where steps have been taken to coordinate the development of university and college programmes.

This has led to a characteristic pattern which has been observed in many agencies. Typically, an individual with some background in computing will hear of or see a geographical information system or automatic cartography system in operation and, by attending conferences or workshops, become a promoter within his own agency. With luck he will eventually assemble sufficient resources to acquire a system, and can expect to be named manager of it. The rest of the agency will be very happy to know that the group is involved in the new technology, but equally happy to know that all obligation to understand it rests with the resident expert.

The only effective way out of this sociological impasse appears to be a complete and comprehensive study of functional requirements, with the full cooperation of the director of the agency. The ways in which the products of a GIS will be used by the agency must be documented well before the GIS is acquired and installed, so that they become obligations on the agency staff to use and understand the function of the system.

The only possible alternative would be to ensure that all staff were introduced to GIS as part of their basic education, whatever their application area: if this were possible, it would take decades to achieve. In short, the greatest obstacle to the greater use of GIS will continue to be the human problem of introducing a new technology which requires not only a new way of doing things, but also has as its main purpose permitting the agency to do many things which it has not done before and often does not understand.

Extensive use of digital approaches to handling geographical data draws attention to a number of technical issues which are often transparent or ignored in manual processing. It is necessary to be explicit about such formerly implicit issues as accuracy, precision and generalization. Coordinate systems, projections and transformations must be specified precisely, because there are no globally accepted standards. Those involved in handling geographical data in North America will continue to suffer through a complex maze of different coordinate systems, viz., Universal Transverse Mercator and latitude/longitude at national scales, but a variety of systems at state and provincial levels.

Several developments in the next few years are likely to affect the general level of user awareness of GIS. Although the first text on GIS has only recently appeared (Burrough 1986), the set of teaching materials is likely to grow rapidly in response to obvious needs. New journals are appearing. Several universities have instituted programmes at undergraduate and graduate levels, and there are technical diploma programmes available at several institutions. The National Science Foundation has proposed a National Center for Geographic Information Analysis, which it will fund in part and which could have a major coordinating role. But despite this activity, the organizations, journals and textbooks, and there has been no systematic approach to GIS in the educational system. The term is encountered in courses in departments of geography, surveying and forestry, in other words primarily in applications of GIS, by students with little or no technical background. There have been attempts in North America in various sectors to coordinate the development of specialist programmes, but these will not be successful as long as they are perceived as benefiting one sector at the expense of another. The federal governments are unlikely to have much influence on the educational system until they can coordinate their own interest in geographical data handling, and there is little sign of their doing so at present. The greatest progress has been made by provincial governments in Canada, where steps have been taken to coordinate the development of university and college programmes.

This has led to a characteristic pattern which has been observed in many agencies. Typically, an individual with some background in computing will hear of or see a geographical information system or automatic cartography system in operation and, by attending conferences or workshops, become a promoter within his own agency. With luck he will eventually assemble sufficient resources to acquire a system, and can expect to be named manager of it. The rest of the agency will be very happy to know that the group is involved in the new technology, but equally happy to know that all obligation to understand it rests with the resident expert.

The only effective way out of this sociological impasse appears to be a complete and comprehensive study of functional requirements, with the full cooperation of the director of the agency. The ways in which the products of a GIS will be used by the agency must be documented well before the GIS is acquired and installed, so that they become obligations on the agency staff to use and understand the function of the system.

The only possible alternative would be to ensure that all staff were introduced to GIS as part of their basic education, whatever their application area: if this were possible, it would take decades to achieve. In short, the greatest obstacle to the greater use of GIS will continue to be the human problem of introducing a new technology which requires not only a new way of doing things, but also has as its main purpose permitting the agency to do many things which it has not done before and often does not understand.

Extensive use of digital approaches to handling geographical data draws attention to a number of technical issues which are often transparent or ignored in manual processing. It is necessary to be explicit about such formerly implicit issues as accuracy, precision and generalization. Coordinate systems, projections and transformations must be specified precisely, because there are no globally accepted standards. Those involved in handling geographical data in North America will continue to suffer through a complex maze of different coordinate systems, viz., Universal Transverse Mercator and latitude/longitude at national scales, but a variety of systems at state and provincial levels.

Several developments in the next few years are likely to affect the general level of user awareness of GIS. Although the first text on GIS has only recently appeared (Burrough 1986), the set of teaching materials is likely to grow rapidly in response to obvious needs. New journals are appearing. Several universities have instituted programmes at undergraduate and graduate levels, and there are technical diploma programmes available at several institutions. The National Science Foundation has proposed a National Center for Geographic Information Analysis, which it will fund in part and which could have a major coordinating role. But despite this activity, the supply of qualified Ph.D.s in GIS to staff these new programmes and implement the research which will allow the field to develop remains woefully inadequate to meet current demand.

4. Spatial data sets

Many government agencies have undertaken surveys of existing spatial data sets in order to improve access and reduce duplication. Notable examples are the USGS in the late 1970s, EMR in 1984 and Ontario and Quebec in the early 1980s. In some cases these include only data sets where the locational identifier is specific, in the form of a coordinate pair; in others the locational identifier can be nominal: a pointer to a location specific file, as in the case of a street address. Some of these studies have provided directories only, but others have gone on to estimate existing and potential usage (e.g. the EMR study). In the latter case the results were almost uniformly disappointing: where data sets had been compiled at least partly for use by others, the level of use has almost always been overestimated.

In general, where such studies have been made at the federal level they have been undertaken by a single agency with little prospect of continuing update. On the other hand, provincial governments, such as that in Ontario, have been more successful at undertaking coordinated, multi-agency surveys, typically by a Cabinet decision assigning responsibility to one department (Natural Resources in the case of Ontario).

There seems little prospect of coordination of key spatial data sets in North America because of problems of confidentiality and the division of responsibility between the three levels of governments. In Canada, the administrative records from Revenue Canada have been linked to postcodes and Statistics Canada has linked postcodes to census areas; but the detailed network of postcode boundaries is not planar and has not been digitized, and there are changes in the lowest level of census areas in each census. Files of land parcels are a municipal responsibility in most provinces and are linked to the system of boundaries for municipal elections, which do not respect any of the federal units. Because of these problems there has been a move to the concept of municipal census in some cities: certainly the trend in North America is for a decrease in the power of central governments to coordinate the gathering of socio-economic data.

Several general points can be made based on knowledge of the development and use of digital spatial data sets in both Canada and the U.S. Firstly, the potential for exchange and common use of digital cartographic data in both countries is not yet being realized. Although the data structures in use in various systems are broadly compatible, there is little or no standardization in detail, and each of the major systems currently in use has developed independently to satisfy the specific needs of individual agencies. Problems of coordinate systems, scales and projections, lack of edgematching and incompatible use of tiling and framing remain major barriers to exchange and sharing, and encourage duplication of effort. Efforts which have been made to standardize formats, such as the U.S. Geological Survey's Digital Line Graphics (DLG) and the Canadian Standard Data Transfer Format (SDTF), have had little effect, whereas there has been rather more *de facto* standardization because of the widespread installation of systems from the same vendor; Intergraph's Standard Interchange Format (SIF) has become quite successful for this reason.

The option of making digital data rather than buying them or copying them from another system will remain popular as more and more vendors enter the market with unique data formats. But although the incentive to go digital will grow as costs of software and hardware drop and as functionality improves, this will not immediately affect the amount of data sharing, since it will reduce not only the cost of buying or importing data, but also the entry costs for hardware and software related to digitizing them locally. It is likely, therefore, that duplication will remain common for some time to come. Sharing is also hindered by lack of information on what is available and by lack of rapid access, and no fundamental change is likely in either factor in the near future.

In summary, the author expects that, in principle, the ready availability of data on a wide range of geographical themes should be one of the major benefits of the trend toward digital cartography and GIS. In practice the picture in 1987 is one of very incomplete coverage both regionally and thematically, a situation owing more to historical and behavioural factors than to design. Each major system has developed independently and its designers have felt little pressure in the past to standardize or to establish the capabilities for the interchange of data. As a result, very little transfer of data takes place, although there is more frequent use of another agency's analytic facilities. At the same time, in the absence of coordination, digital capabilities are being acquired at a rapid rate by all types of agencies, and the popularity of a small selection of systems is creating a situation of informal and arbitrary standardization.

5. Costs and benefits

There has been very little in the way of formal accounting of GIS operations, largely because, in most agencies, they are still regarded as experimental. The only GIS known to the author which has been in operation for sufficiently long with stable hardware and software to permit any kind of realistic accounting is CGIS, which has kept accurate and detailed records of its own operational costs for some time (Goodchild and Rizzo 1987). Experience in the software engineering field in the U.S. indicates that costs of such record keeping would run to about two to three per cent of the total cost of a project.

Several agencies have undertaken some form of bench-mark testing as part of their process of acquiring a GIS, although others have relied entirely on the claims of vendors, in some cases with disastrous results. However, bench-mark testing has almost always been designed to test the existence of functions, rather than to test the speed of performance and therefore costs of performance. This deficiency merely reflects the immature state of the GIS software industry and will presumably change slowly over time as more and more agencies acquiring GIS use bench-mark tests to determine the extent to which the system being offered will or will not perform the required workload within the prescribed schedule. Estimation of workload will also become increasingly important if a significant market develops for GIS services as distinct from sales of systems.

The unit of output of a GIS is a processed information product, in the form of a map, table or list. To assess the cost of this product would require a complex set of rules for determining cost accrual: these are fairly straightforward in the case of the direct costs of the product, but less so for the capital costs of the system and the costs of inputting the data sets from which the product was derived. We are not aware of any system which has attempted to establish such a set of rules.

Far more difficult is the determination of the benefits of the product. The benefit of a unit of output can be defined as an item of information, by comparing the eventual outcome of decisions made using it, and decisions made without it. In almost all circumstances, both such outcomes would have to be estimated. The author's discussion of this issue in relation to the U.S. Forest Service supposed that it might be possible to identify a number of decisions in which some of these problems could be resolved, but that it would be virtually impossible to compute total benefit. The proposed solution was to institute a programme of identifying and tracking suitable decisions, in order to form the basis for a largely qualitative evaluation of benefit.

In those cases where a GIS product replaces one produced manually, it is relatively easy to resolve the cost/benefit issue by comparing the costs of the two methods: the benefits are presumably the same. Where this has been done, as for example in a study of the functional requirements of the Ontario Forest Resource Inventory, the option of using a GIS is usually cheaper by an order of magnitude. However, this approach glosses over the fact that many of the products would not have been either requested or generated if the GIS was not available.

These problems are not unique to GIS, but occur in similar fashion in all information systems and in many other applications of data processing. It is clear that, without a great deal more experience of production, the case for a GIS will not be made on the basis of a direct, quantitative evaluation of tangible benefits, but must include less direct and more subjective and better informed decision-making and an improved quality of data.

Nor does the author know of any direct attempt to compare the costs and benefits of an installed operating GIS with those estimated in planning the installation. There is, of course, abundant anecdotal information of this kind, but little is of much substance. The practice of comprehensive planning for the acquisition of GIS has developed only in the past 5 years or so, and it is still relatively unusual to find that any attempt has been made to estimate costs or benefits in advance. However, it is likely that this sort of evaluation will be required for some of the major acquisitions pending, and this will provide an opportunity to conduct an objective evaluation for the first time.

6. Technical developments

6.1. *Digitizing*

The suspicion has already been voiced that future technical developments will not lead to any marked improvements in methods of manual digitizing. There is, however, considerable room for improvement in automated procedures, and consequent reductions in costs of inputting data. More sensitive scanners will allow greater discrimination between data and noise. Better software, with properties verging on artificial intelligence, will produce more accurate vectorization and identification of features. Rather than employing direct scanning, the systems of the future will probably scan the document in a simple raster and rely on processing to track features logically rather than physically, because memory is becoming less and less a constraint. There are now good raster edit systems on the market and future systems will optimize the relative use of raster and vector approaches to minimize overall time and cost of editing. Finally, better hardware and software will allow greater use of raw documents and there will be less need for the preparation of specialized documents and scribing. In the long term, there is considerable scope for redesigning cartographic documents to make scanning easier and more accurate, for example in the use of special fluorescent inks and bar codes.

These developments may mean that raster input will eventually replace vector digitizing for complex documents and the input of bulk data. Manual digitizing will remain an option with a low capital cost for the input of simple documents and for systems requiring the input of low volumes of data.

6.2. *Raster versus vector*

In most of the sectors listed in this paper, the current market is dominated by vector rather than raster systems. Of course, all systems are to some extent hybrids and the distinction refers to the form of storage used for the bulk of the data base, and to the data structure used in most data processing in the system.

The primary concern in choosing between raster and vector systems is the nature of the data in that sector, and in some sectors it has taken years of experiment to resolve the issue. With land parcels the need to work with arbitrarily shaped but precisely located property units has led to vector systems which allow the definition of entities that are linked to statutory responsibilities. The nature of elevation data suggests that raster storage be used and most satellite-based imagery originates in that form. These conclusions are fundamental to the nature of each sector and are unlikely to change quickly.

On the other hand, any vendor wishing to market a comprehensive system whose applicability spans several sectors must consider both types of data and transfers between them. Several vector systems (e.g. ESRI, Intergraph) offer raster-based functionality for handling topographic data. Ultimately, all comprehensive GIS will recognize four locational primitives, viz., point, line, area and pixel, and will permit appropriate processes on all four. There will also continue to be systems designed primarily for image processing which retain the raster as the central mode of storing data, but permit vector features for specific functions.

7. Sources of research and development funding

In North America, no government agency has a general mandate to fund research and development in geographical data handling: all those agencies which do fund have specialized needs in defined applications or a general mandate to fund basic research. There is no indication that this will change in the next 5 to 10 years.

The state of funding thus depends very much on the state of the responsible agencies in each sector. At present, public attention on environmental issues in North America is perhaps less focused than before and concerns about forestry, oil and gas are probably past their recent peaks. On the other hand, substantial amounts of money are being allocated to research into water and some interest in GIS applications in this field can therefore be expected. There is considerable interest in military applications, due partly to the general revival of funding in this sector and partly to the needs of new weapons systems, and this interest can be expected to continue for some time and to lead to research and development in topographic applications of GIS technology, particularly in methods of editing, verifying and updating dense digital elevation models (DEM). Some of the proposed work in artificial intelligence (AI) for the Strategic Defence Initiative (SDI) will have relevance to GIS because of its concern for image processing, and there may be other connections between GIS and strategic weapons systems.

Marketing and related applications were earlier identified as a major growth area. Research and development in such applications are likely to be funded by the private sector. However, the field has intersections with a number of governmental and municipal activities, particularly statistics, public works, crown corporations in transport, education and management of emergency services, and small amounts of funding may find their way into GIS research and development through such channels. There is also an increasing tendency for agencies for funding basic research, such as the National Science Foundations, the Natural Sciences, Engineering and

Research Council of Canada and the Social Sciences and Humanities Research Council of Canada, to favour programmes which match funds raised in the private sector.

8. Nature of GIS research

Over the past 20 years the roles of GIS, automated cartography and computer-aided design (CAD) have frequently been confused, both in the relative applicability of each technology in various fields and in the direction of basic research. The respective data structures have very little in common: the literature of GIS makes very little reference to automated cartography or to computer-aided design, and whole topic areas which are intimately related to GIS, such as spatial analysis and spatial statistics, have no relevance to automatic cartography or to CAD. On the other hand, the development of technology for digitizing and display has clearly benefited from the influence of the much larger market for CAD.

The author believes that GIS is a unique field with its own set of research problems, although the entire GIS community would probably not agree with this view. A GIS is a tool for manipulation and analysis of spatial data; it therefore stands in the same relationship to spatial analysis as standard statistical packages such as SAS and SPSS stand to statistical analysis. This is radically different from the purpose of CAD or automated cartography, and has led to the development of fundamentally different data structures and approaches.

It follows that the set of potential applications of GIS is enormous, and is not satisfied by any other type of software. Future developments in GIS will depend on better algorithms and data structures, and continuing improvements in hardware. But they also need research in spatial analysis, in the development of better methods of manipulating and analysing spatial data, and towards a better understanding of the nature of spatial data themselves through such issues as generalization, accuracy and error. Thus, research in GIS needs to be concentrated in three areas: data structures and algorithms, spatial analysis and spatial statistics. Development of the hardware will probably continue to be motivated by larger markets in computer graphics and CAD.

9. Lessons learned

A number of lessons can be learned from this experience. Firstly, North American political structures make it difficult to organize coordinated approaches to the development of technologies such as GIS. In Canada, the greatest success has been at the provincial level, but even here there are consistent patterns of duplication and lack of planning. The North American suspicion of big government and bureaucracy also operates against attempts to establish large, well-coordinated spatial data bases. Mapping is split over many agencies and over several levels of government. The result has been a haphazard approach in which such standards as have emerged have come about because of the small number of vendors in the market, rather than through any central coordination.

Secondly, there is an increasing gap between the need for qualified staff in agencies which have acquired GIS and the ability of the educational system to provide them. Again, this gap stems partly from the lack of central planning in the educational system and the level of autonomy enjoyed by most educational institutions. A process of natural selection is occurring, among both institutions and disciplines, and it is too early to tell what the result will be.

Thirdly, GIS is a highly attractive technology and suffers from the problems of all such technologies. It has repeatedly been wrongly sold as a solution in response to

180

needs which were poorly defined or not defined at all, and to clients who did not really understand its capabilities or limitations. Many failures have resulted from the acquisition of the wrong type of system because of poor advice or lack of advice. The success of the technology is in many ways dependent on the availability of good advice, through either centres of excellence in the public sector or competent consultants in the private sector.

Fourthly, many failures in North America have resulted from agencies attempting to develop their own systems with inadequate resources and exaggerated goals. The systems currently on the market only partially satisfy the notion of a GIS and yet typically contain ten or more functions, and represent investments of several man-years of programmers' time. Yet there have been many examples of agencies embarking on projects to develop similar functions from scratch. Such in-house R. & D. in the public sector would be of some benefit to society at large if it resulted in substantial cooperation with the private sector and eventual sales, but the record of such cooperation in North America has not been good: instead, developments in the public sector have tended to compete with the private sector, with all the attendant problems of hidden subsidies and unsatisfactory procedures for tendering.

Finally, there are just as many problems, and possibly more, on the management side of implementing an information system as there are on the technical side. A primary benefit of a GIS lies in the new capabilities which it introduces, rather than in the ways in which it allows old tasks to be done more efficiently or more cheaply. To be successful, then, a GIS requires strong and consistent motivation on the part of all users, the great majority of whom will have no technical understanding of the system. Such motivation will not occur naturally, however user-friendly the system, however impressive its products, and however sophisticated its functionality. It is absolutely essential that the users of the system be the ones who plan it and arrange for it to be acquired. Systems which have been installed on a trial or experimental basis, 'to allow potential users to see the benefits', almost invariably fail once the honeymoon is over.

10. Conclusions

A successful programme in GIS at the national scale would seem to need the coordination of five types of effort. Firstly, it would need a set of valid applications. On the basis of experience in North America there seems at the moment no reason to doubt that such areas exist and that they will grow rapidly in the near future. Secondly, it requires a set of active vendors. Again, there is no reason to doubt that the products available will grow in number or become more sophisticated in the next few years.

The size of any nation's share of the international pie is likely to depend partly on the size of the nation's application areas, but primarily on the remaining three types of effort. The third is the educational sector, which will provide trained personnel to run systems, conduct basic research, staff the vendors and train future generations. Although various governments in North America have funded programmes in various universities and institutes, the total 'top-down' influence remains small and most courses and programmes exist because of a 'bottom-up' perception of a demand for graduates.

The fourth type of effort is in research and development. It seems from past experience that most major breakthroughs in research will occur in government establishments and in universities, rather than in the private sector: commercial incentives appear to have led to success in adapting and improving, but not to major new directions. The most significant advances in the GIS field remain those originally made in CGIS and subsequently at the Harvard Laboratory for Computer Graphics and Spatial Analysis. Governments will, therefore, remain the major sources of funding, and the major control on the rate at which the field progresses.

Finally, because of the managerial issues discussed earlier, successful applications of GIS require not only the four types of effort already mentioned, but also the existence of a substantial source of expertise which is independent of vendors and acquiring agencies and is capable of mediating the process of acquisition and ensuring that it leads ultimately to success. This role can be filled by an independent, commercial consulting sector, but it is unlikely that this sector could exist in the early, critical period of a national involvement in GIS. It is role which could be filled by a government centre of expertise, which would have to rely to some extent on imported talent. It is something which has been critically lacking in developing countries and in industrial countries which have come into the GIS field rather late because of the small size of the local market.

Within North America, the successful establishment of the National Center for Geographic Information Analysis which the National Science Foundation has proposed and completion of the selective digitization of the USGS 1:100000 series maps for the 48 contiguous states, an essential component of the Bureau of the Census' plans for the preparation of maps for the 1990 census of population, may be expected to have a significant impact on the development of GIS. Both projects have aroused considerable interest, the first among researchers and the second among commercial agencies, especially those concerned with marketing and market research. Whether that expectation is fulfilled is another matter; prognostication in a field developing so rapidly is a hazardous enterprise.

Acknowledgment

This paper is based on a report produced by Tomlinson Associates Ltd and published within the Chorley report (Committee of Enquiry 1987).

References

BURROUGH, P. A., 1986, *Principles of Geographical Information Systems for Land Resources Assessment* (Oxford: Oxford University Press).

COMMITTEE OF ENQUIRY CHAIRED BY LORD CHORLEY, 1987, *Handling of Geographic Information* (London: HMSO), Appendix 6.

GOODCHILD, M. F., and RIZZO, B. R., 1987, Performance evaluation and work-load estimation for geographic information systems. *International Journal of Geographical Information Systems*, **1,** 67.

RAKER, D. S., 1982, Computer mapping in geographic information systems for market research. *Proceedings, National Computer Graphics Association*, **2,** 925.

SECTION 6

GIS Applications

Overview

This section provides some recent examples of applications of GIS. In the first article, Welch and others demonstrate how a micro-computer GIS was used in aquatic resource evaluation. The second paper by Jensen and Christensen describes the development of a GIS model to assist in waste disposal site selection. In the last paper, Usery and others discuss an approach to using a knowledge-based GIS for a geological engineering application.

Suggested Additional Reading

Merchant, J.W. (ed.) 1988. Special GIS Issue. *Photogrammetric Engineering and Remote Sensing*. 54(11):1485–1650.

Merchant, J.W. and W.J. Ripple (ed.) 1987. Special GIS Issue. *Photogrammetric Engineering and Remote Sensing*. 53(10):1309–1492.

Remote Sensing and Geographic Information System Techniques for Aquatic Resource Evaluation*

R. Welch and *M. Madden Remillard*
Laboratory for Remote Sensing and Mapping Science, University of Georgia, Athens, GA 30602
R. B. Slack
U. S. Environmental Protection Agency, Atlanta, GA 30365

ABSTRACT: The spread of aquatic plants in Lake Marion, South Carolina necessitated an assessment of the trends in vegetation growth and water quality. Aquatic vegetation maps at 1:10,000 and 1:24,000 scale were produced by photogrammetric techniques from color infrared aerial photographs recorded on six dates between 1972 and 1985. These and other vector map products depicting bathymetry and herbicide applications were converted to raster format (25-m grid cells) to form a cartographic database for the 170 km² study area. Statistical data on nutrients, dissolved oxygen, biological oxygen demand, and turbidity obtained from South Carolina Department of Health and Environmental Control and U.S. Environmental Protection Agency records were also input to the database. A PC-based GIS was then used to relate macrophyte distributions to environmental factors influencing aquatic plant growth. The procedures employed represent an inexpensive approach that can be applied to other resource management tasks.

INTRODUCTION

WITH THE EVOLUTION of geographic information systems (GIS) technology from an experimental to operational mode, increased recognition is being given to the potential for integrating remote sensing and database methodologies to monitor natural resources. An example of this integration process is the construction of a database from aerial photographic, map, and statistical information, and the development of a lake management information system (LMIS) for monitoring aquatic macrophytes and water quality in the large inland reservoirs of South Carolina.

Concern for the future of major lakes and reservoirs within South Carolina prompted the Department of Health and Environmental Control (DHEC) to instigate a comprehensive water quality study of the Santee-Cooper River Basin, the second largest river basin on the East Coast of the United States. The initial focus of the study was on the relationships between the distribution of aquatic macrophytes and water quality in Lake Marion, a reservoir of some 45,000 ha. A major question has been whether changes in water quality will cause an increase in the already extensive aquatic plant population of upper Lake Marion and promote the spread of undesirable macrophytes to other parts of the system. In order to address that question, DHEC, with support from the U.S. Environmental Protection Agency (EPA), contracted with the Laboratory for Remote Sensing and Mapping Science (LRMS), University of Georgia, to inventory the aquatic macrophytes in upper Lake Marion, determine changes over time, and develop a GIS database that would allow these changes to be related to water quality, bathymetry, and sedimentation (Welch *et al.*, 1985 and 1986; Welch and Remillard, 1986). The methodology used to create an integrated database and develop a prototype LMIS suitable for aquatic resource evaluations is the subject of this paper.

STUDY AREA

Lake Marion was formed when the U.S. Army Corps of Engineers impounded the Santee River in 1941 to provide hy-

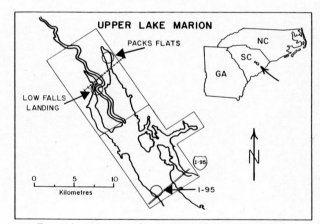

FIG. 1. Location map for the Lake Marion study area.

droelectric power in South Carolina. Although large, the lake is relatively shallow and supports a diverse fish and waterfowl population, making it a popular recreational area.

The specific study area is confined to upper Lake Marion, which extends approximately 23 km northwest of highway I-95 (Figure 1, Plate 1). This 17,000-ha area represents a gradual transition from the alluvial floodplain of the Santee River to the impounded lake (Harvey *et al.*, 1987). Physical characteristics include relatively stable water levels, shallow depths (not exceeding 20 m and averaging less than 3 m), and high turbidity (Patterson and Cooney, 1986). Sediment from the Santee River inflow is deposited in upper Lake Marion as the water velocity of stream flow decreases. The presence of aquatic macrophytes in the lake further encourages deposition which, in turn, creates a favorable habitat for aquatic plant growth.

Aquatic macrophytes are grouped by structure as emergent (rooted plants with leaves extending above or floating on the water surface), submergent (rooted plants growing below the surface), or free floating (non-rooted, surface floating plants). Common emergents found in upper Lake Marion include water primrose (*Ludwigia uruguayensis*), yellow lotus (*Nelumbo lutea*), fragrant water lily (*Nymphaea odorata*), and dollar bonnet (*Bra-*

*Presented at the ISPRS Commission IV meeting during the ASPRS/ACSM Annual Convention, Baltimore, Maryland, March 1987.

PHOTOGRAMMETRIC ENGINEERING AND REMOTE SENSING,
Vol. 54, No. 2, February 1988, pp. 177–185.

0099-1112/88/5402–177$02.25/0
©1988 American Society for Photogrammetry
and Remote Sensing

PLATE 1. High altitude color infrared aerial photograph of upper Lake Marion recorded in 1972.

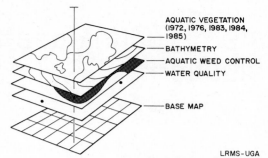

LRMS-UGA
FIG. 2. Data sets in the Lake Marion GIS database.

Landsat Thematic Mapper (TM) image data also have been used to map aquatic vegetation (Jensen *et al.,* 1986). However, the relatively poor spatial resolution of the satellite data and the difficulty of obtaining cloud-free coverage during the growing season limit its use to general classification tasks. High resolution airborne multispectral scanner data are also an option, but suffer severe geometric distortions and are costly to acquire. Because of these problems, large scale color infrared aerial photographs were determined to be the most useful and cost effective source materials for mapping aquatic plant species distributions.

DATABASE CONSTRUCTION

A database of biological and physical characteristics for upper Lake Marion is being constructed for use in an LMIS. This database will be employed to assess the ecological relationships between aquatic macrophytes, water quality, and environmental factors.

The layers in the database include maps of aquatic macrophyte distributions prepared from aerial photographs; a bathymetric map; maps depicting areas in which herbicides have been applied since 1982; and statistical data on water quality (Figure 2). Each of these layers is discussed in the following paragraphs.

AQUATIC MACROPHYTE DISTRIBUTION MAPS

Maps of the distribution of aquatic macrophytes were developed from color infrared aerial photographs in film transparency format recorded in 1972, 1976, 1983, 1984, and 1985 (Table 1). Photos acquired since 1983 have been of large scale (1:8,000 to 1:12,000), facilitating the identification of species.

The photographs were interpreted under high magnification with Bausch and Lomb Zoom 70 and SIS 95 instruments, and polygons representing the different types of aquatic vegetation were delineated on clear polyester overlays registered to the aerial photographs. Each polygon was classified according to the type of vegetation: (1) emergent, (2) submergent, or (3) free floating. Individual emergent species were also identified and verified with field data. Submergents, although readily identifiable by type on the photographs to depths of approximately 3 to 4 m, could not be differentiated by species because of their relatively uniform tone and texture. Therefore, only in those areas for which field data existed were submergents labeled by species.

Table 2 summarizes the photo characteristics of aquatic macrophytes commonly found in Lake Marion. Height and texture parameters are similar to those used by Seher and Tueller (1973) in their color infrared photographic key of marsh vegetation. Height categories include floating (leaves floating on the water surface), low (plants extending up to 15 cm above

senia schreberi). Submergent species commonly mixed with the emergents include Brazilian elodea (*Egeria densa*), hydrilla (*Hydrilla verticillata*), coontail (*Ceratophyllum demersum*), Southern naiad (*Najas minor*), and pondweed (*Potamogeton* spp.). Duckweed (*Lemna perpusilla* and *Spirodela* spp.), water-fern (*Azolla caroliniana*), and water-meal (*Wolffia papulifera*) are the dominant free floating species. Taxonomic classification is according to Radford *et al.* (1983).

Previous studies of macrophyte distributions in upper Lake Marion date back to 1977 when general aquatic vegetation maps were produced by the U.S. Army Corps of Engineers Waterways Experiment Station from small format color infrared aerial photographs (Link and Long, 1978). More recently, DHEC performed an extensive field survey to map aquatic macrophytes using a Motorola Miniranger III Automated Positioning System (APS) to record boundary coordinates of aquatic plant populations (Harvey *et al.,* 1983). The APS is labor intensive and limited to relatively open water areas where trees do not block the transmitted signals. Because nearly 65 percent of upper Lake Marion is a dense cypress-tupelo swamp, a more efficient mapping technique was required.

	1972	1976	1983	1984	1985 (June)	1985 (Sept)
Date of Acquisition	9/22/72	11/18/76	9/8/83	9/6/84-9/7/84	6/9/85	9/14/85-10/7/85
Nominal Scale	1:128,000	1:20,000	1:10,000	1:28,000	1:12,000	1:12,000
Flying Height AGL (m)	19,500	3,050	1,525	1,220	1,830	1,830
Camera	Wild RC-10	Fairchild KA-30	Zeiss RMK A 15/23	Wild RC-10	Wild RC-10	Wild RC-10
Total Number of Photos	2	110	170	270	105	139

the surface), medium (15 cm to 1 m), and tall (greater than 1 m). Texture is specified as fine, medium or coarse.

Aquatic macrophyte distribution maps, based on interpretation of the aerial photographs, were produced at 1:10,000 and 1:24,000 scale for 1983, 1984, and 1985, and at 1:24,000 scale for 1972 and 1976. The following method was employed to construct the maps.

A framework of ground control points (GCPs) was identified on a set of 1:28,000-scale color infrared photographs of the study area and their UTM coordinates were digitized from existing 1:24,000-scale USGS topographic maps and transferred to gridded planimetric base sheets of 1:5,000 scale. Numerous additional (pass) points common to the larger-scale photographs on which the polygons were delineated were also annotated on the 1:28,000-scale photographs. The pass points were then transferred to 1:5,000-scale planimetric base sheets using a Bausch and Lomb

TABLE 2. PHOTO CHARACTERISTICS OF AQUATIC MACROPHYTES

Aquatic Macrophyte	Color	Height	Texture
EMERGENT SPECIES			
Dollar bonnet	white to pale pink	floating	fine
Lotus	light to bright pink	floating to medium	fine to medium
Primrose	purple to pink	low to medium	medium
Water lily	white to pale pink	floating to low	fine
Unidentified Emergent	white to bright pink	floating to medium	fine to medium
SUBMERGENT SPECIES			
	Dark blue to black	subsurface to floating	fine
FREE FLOATING SPECIES			
Duckweed	white	floating	fine
MIXED SPECIES			
Emergents Dominant	pink to dark pink	floating to medium	medium
Submergents Dominant	blue to black	subsurface to low	fine to medium
Free Floating Dominant	white to pale pink	floating to low	fine to medium

ZT-4 Zoom Transfer Scope to provide a control network for the large-scale photographs. In a final step, the macrophyte polygons were transferred from the large-scale photographs to the base sheets with the aid of the Zoom Transfer Scope.

The base sheets were then photographically reduced by a factor of two and mosaicked together to permit the compilation of maps of 1:10,000 scale for the entire study area. Where necessary, the polygons were generalized to achieve a minimum size mapping unit of 0.01 ha (10 by 10 m).

The completed 1:10,000-scale macrophyte distribution maps were also photographically reduced to 1:24,000 scale to allow registration with existing USGS quadrangles (Figures 3a and 3b). Maps depicting changes in aquatic macrophyte distributions were produced by registering maps of different dates to one another and delineating areas of change in macrophyte classification (Figure 3c).

BATHYMETRIC MAP

A bathymetric map of Lake Marion produced from fathometer recordings taken from boat surveys along transects of the lake was obtained from the U.S. Geological Survey (USGS, 1984). Contours depicted on the map at a 0.6-m interval were adjusted to compensate for the lower summer pool water level at the time the aerial photographs were recorded. This was accomplished by raising all depth contours by one interval or 0.6 m.

HERBICIDE APPLICATION MAPS

Aquatic plant management practices in Lake Marion are directed by the South Carolina Water Resources Commission (WRC) in conjunction with the South Carolina Aquatic Plant Management Council. Specific areas of the lake are sprayed annually with herbicides, such as Diquat, Aquathol-K, and Sonar, to control excessive macrophyte growth. The single largest control effort in South Carolina in 1986 was conducted in upper Lake Marion at a cost of approximately $460,000 (South Carolina Aquatic Plant Management Society Newsletter, 1986). Based on information provided by DHEC and WRC, maps depicting the area of herbicide application were prepared for 1982, 1983, 1984, and 1985 (Figure 4).

WATER QUALITY STATISTICAL DATA

The WRC and DHEC monitor water quality in Lake Marion. Long term sampling stations in upper Lake Marion cluster in

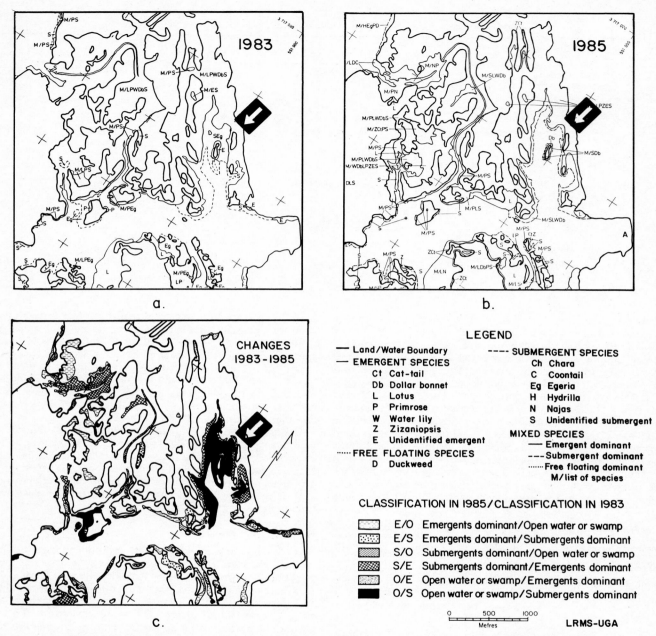

a.

b.

c.

LEGEND

— Land / Water Boundary
— EMERGENT SPECIES
- Ct Cat-tail
- Db Dollar bonnet
- L Lotus
- P Primrose
- W Water lily
- Z Zizaniopsis
- E Unidentified emergent
····· FREE FLOATING SPECIES
- D Duckweed

---- SUBMERGENT SPECIES
- Ch Chara
- C Coontail
- Eg Egeria
- H Hydrilla
- N Najas
- S Unidentified submergent

MIXED SPECIES
— Emergent dominant
--- Submergent dominant
····· Free floating dominant
M/ list of species

CLASSIFICATION IN 1985 / CLASSIFICATION IN 1983

- E/O Emergents dominant/Open water or swamp
- E/S Emergents dominant/Submergents dominant
- S/O Submergents dominant/Open water or swamp
- S/E Submergents dominant/Emergents dominant
- O/E Open water or swamp/Emergents dominant
- O/S Open water or swamp/Submergents dominant

0   500   1000
Metres   LRMS-UGA

FIG. 3. (a, b) Macrophyte distribution maps for 1983 and 1985, respectively; and (c) Macrophyte change map for 1983 to 1985.

three general areas: I-95 (6 stations), Low Falls (11 stations), and Packs Flats (9 stations) (see Figure 1). These data are stored in the Environmental Protection Agency database, STORET.

Monthly data for April through September (for the 13-year study period) were averaged to obtain an overview of annual growing season trends in nutrients (nitrogen and phosphorus), dissolved oxygen (DO), biological oxygen demand (BOD), and turbidity (Figure 5). These water quality parameters were selected because of their relationships to aquatic macrophyte growth and the availability of consistent records for the study period (Table 3).

In order to create a computer database that could be utilized with the various data sets to assess water quality and the distribution of aquatic macrophytes, all data sets were converted to a raster format compatible with the cell-based GIS software package, pMAP, available from Spatial Information Systems. This

package is designed for use on an IBM PC/AT or compatible computer and contains all the necessary analytical functions required for lake management applications. However, it does

TABLE 3.   SELECTED WATER QUALITY PARAMETERS

Water Quality Parameter	Relation to Aquatic Plant Growth
Nutrients (Nitrogen and Phosphorus)	Essential elements for plant growth
Dissolved Oxygen	Released by plants and required by flora and fauna for respiration
Biological Oxygen Demand	Index for lake eutrophication
Turbidity	Determines light penetration and plant distribution

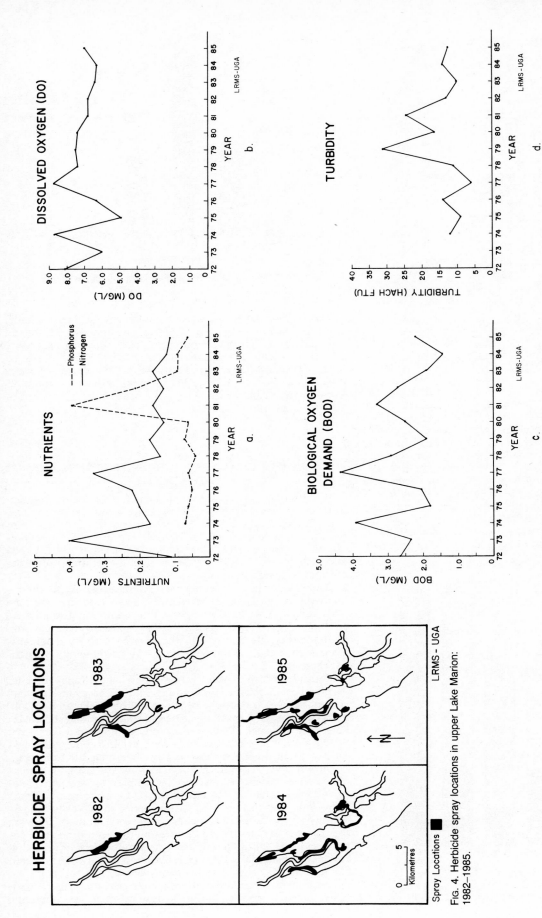

FIG. 5. Annual average growing season levels of water quality parameters: (a) nutrients; (b) dissolved oxygen demand; and (d) turbidity.

FIG. 4. Herbicide spray locations in upper Lake Marion: 1982-1985.

189

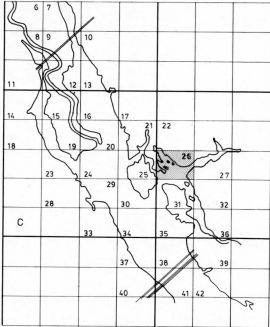

FIG. 6. The upper Lake Marion study area divided into 2-by 2.5-km map segments.

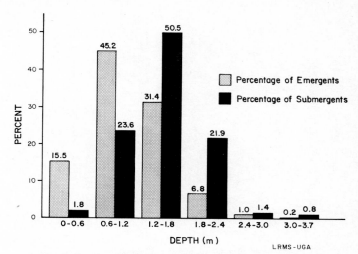

FIG. 8. Depth distributions of emergents and submergents.

suffer two drawbacks: (1) the lack of a good data capture routine that will rasterize digitized map data in the pMAP format; and (2) an affine black-and-white alphanumeric character output that is difficult to interpret. These problems were rectified by adding the Desktop Digitizing Package (DDP), available from R-WEL, Inc., which features programs for the capture (CAPTURE), rasterization (RASTER), and color display (SHOW) of vector data digitized from maps or photographs. The pMAP/DDP software combination provided an inexpensive, easy to use, GIS capability suitable for resource management tasks.

To facilitate the use of pMAP, upper Lake Marion was divided into 2- by 2.5-km map segments keyed to the UTM coordinate

system (Figure 6). The maps were manually digitized by segment and the data converted to pMAP format using the CAPTURE and RASTER routines. To accommodate the pMAP format and accelerate processing of the data, a grid cell size of 25 by 25 m was selected for the database (Berry and Reed, 1987). Comparison of raster data sets at 5-m, 25-m, and 50-m resolution have indicated 25 m to be acceptable for analysis tasks based on source maps of 1:10,000 scale and smaller.

ANALYSIS OF MACROPHYTE DISTRIBUTION AND WATER QUALITY

The database was used to assess changes in the distribution of aquatic macrophytes (Plate 2a). Map segments for any two dates (Plates 2b and 2c) can be subtracted and the changes represented as a color display (Plate 2d). Because each pixel represents an area of 0.0625 ha, a simple computer count of the grid cells in each class provides a quantitative measure of the changes in distribution.

Overall, the total areal extent of aquatic macrophytes remained at about 1,800 ha between 1972 and 1984, although the ratio of emergents to submergents did change significantly. As shown in Figure 7, emergents steadily increased, whereas submergents decreased. In 1985, the area covered by emergents alone exceeded 1,800 ha, with submergents showing no apparent increase.

Macrophyte distributions were mapped before (June) and after (September) herbicide applications in 1985 to evaluate weed control efforts. A decrease of about 140 ha in the area of submergents resulted from the herbicide applications; however, emergent and free floating macrophytes increased by 95 ha, yielding a net loss of 45 ha. The decrease in submergents was most dramatic in those areas specifically targeted for spraying.

The effectiveness of the GIS approach for assessing the impact of herbicides is demonstrated for the sample map segment (Plates 2e and 2f). In Plate 2e, the area of herbicide application is shown surrounded by concentric dispersal zones of 100-m width. By overlaying the herbicide dispersal map with the map showing the changes in macrophyte distribution between June and September (Plate 2d), an integrated map product is produced that reveals a substantial decrease in submergents outward for about 200 m from the original spray area (Plate 2f).

The integrated database approach also may be used to spatially compare aquatic plant growth with other environmental

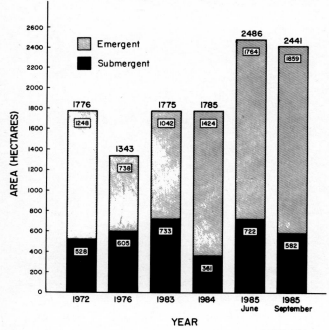

FIG. 7. Total area of aquatic macrophytes: 1972 to 1985.

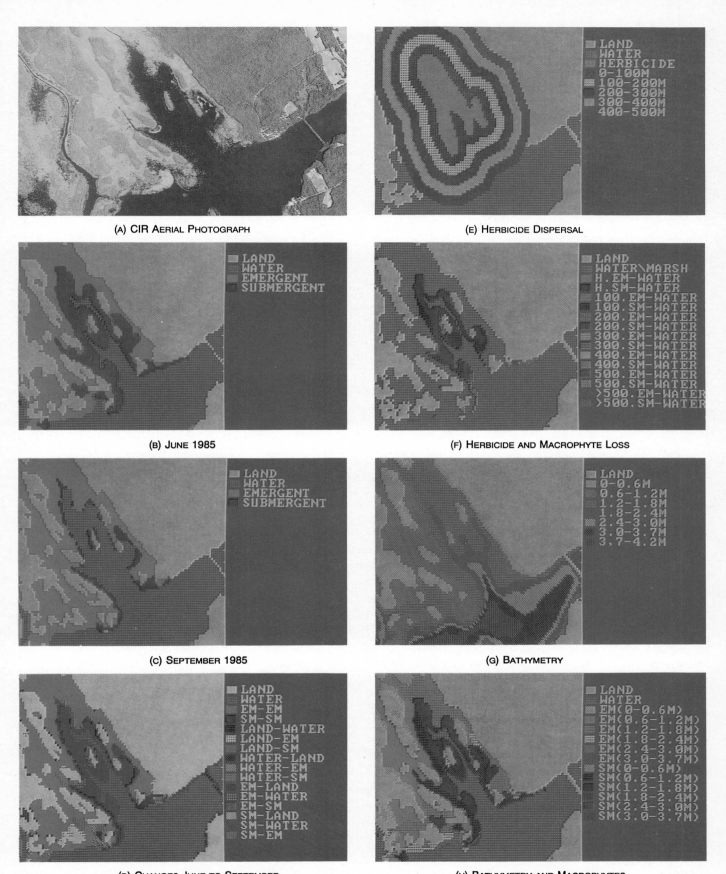

(A) CIR AERIAL PHOTOGRAPH

(E) HERBICIDE DISPERSAL

LAND
WATER
HERBICIDE
0-100M
100-200M
200-300M
300-400M
400-500M

LAND
WATER
EMERGENT
SUBMERGENT

(B) JUNE 1985

LAND
WATER\MARSH
H.EM-WATER
H.SM-WATER
100.EM-WATER
100.SM-WATER
200.EM-WATER
200.SM-WATER
300.EM-WATER
300.SM-WATER
400.EM-WATER
400.SM-WATER
500.EM-WATER
500.SM-WATER
>500.EM-WATER
>500.SM-WATER

(F) HERBICIDE AND MACROPHYTE LOSS

LAND
WATER
EMERGENT
SUBMERGENT

(C) SEPTEMBER 1985

LAND
0-0.6M
0.6-1.2M
1.2-1.8M
1.8-2.4M
2.4-3.0M
3.0-3.7M
3.7-4.2M

(G) BATHYMETRY

LAND
WATER
EM-EM
SM-SM
LAND-WATER
LAND-EM
LAND-SM
WATER-LAND
WATER-EM
WATER-SM
EM-LAND
EM-WATER
EM-SM
SM-LAND
SM-WATER
SM-EM

(D) CHANGES JUNE TO SEPTEMBER

LAND
WATER
EM(0-0.6M)
EM(0.6-1.2M)
EM(1.2-1.8M)
EM(1.8-2.4M)
EM(2.4-3.0M)
EM(3.0-3.7M)
SM(0-0.6M)
SM(0.6-1.2M)
SM(1.2-1.8M)
SM(1.8-2.4M)
SM(2.4-3.0M)
SM(3.0-3.7M)

(H) BATHYMETRY AND MACROPHYTES

PLATE 2. Digital data sets in the lake management information system database: (a) CIR aerial photograph of aquatic macrophytes; (b) and (c) digital map segments of macrophyte distributions for June and September, 1985; (d) changes June to September, 1985; (e) herbicide application, 1985; (f) integrated macrophyte changes and herbicide application; (g) bathymetry; (h) integrated bathymetry and June 1985 macrophytes.

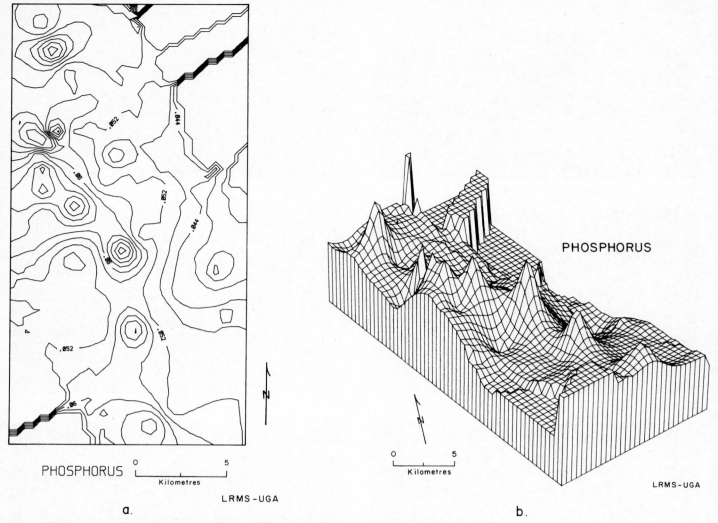

FIG. 9. (a,b) Two- and three-dimensional plots of the distribution of phosphorus in the sample map segment.

factors such as bathymetric data. For example, a map layer for macrophyte distribution in June 1985 (Plate 2b) was overlaid with the bathymetric map (Plate 2g) to establish water depths associated with emergents and submergents (Plate 2h). Based on statistics derived from this composite data set, 92 percent of the emergents were found at depths of 0 to 1.8 m, and 96 percent of the submergents grew at depths of 0.6 to 2.4 m (Figure 8). The maximum depth of macrophyte detection was between 3.0 and 3.7 m for both emergents and submergents. Plants growing at depths greater than 3 to 4 m may not be detectable on color infrared aerial photographs (Martyn et al., 1986). Despite this limitation, the data on macrophyte growth and water depth provide a basis for herbicide selection and application.

Preliminary results of a spatial comparison of macrophyte distributions and water quality indicate that between 1972 and 1985 fluctuations in annual average growing season levels of nitrogen, DO, BOD, and turbidity over the entire study area have not varied significantly and apparently are not related to growth trends noted in upper Lake Marion. Phosphorus levels, however, increased sharply between 1980 and 1983 (see Figure 5a). Because aquatic plants absorb nutrients in part through their foliage, and tend to absorb excessive amounts of essential nutrients (such as phosphorus) that are normally present in low concentrations, the increase in phosphorus may have increased emergent macrophyte growth after 1983 (Figures 9a and 9b)

(Barko et al., 1986). Submergent growth, also expected to be enhanced by high phosphorus levels, may have been effectively checked by herbicide spraying.

CONCLUSION

The integration of remote sensing and database technologies allowed the utilization of a GIS approach to monitor water quality and the growth of aquatic macrophytes between 1972 and 1985 for a study area of some 17,000 ha. The procedures employed for this study can be extended to most types of resource inventories, and can be conducted with a suitably equipped IBM PC/AT or compatible machine and inexpensive GIS software designed to work with databases in raster format. Resource managers thus have an alternative to costly minicomputer based GIS systems for inventory tasks.

ACKNOWLEDGMENTS

This study was sponsored in part by the Environmental Protection Agency (Contract #5R- 1301-NAEX), the South Carolina DHEC (Contract # EQ-5-427), and Lockheed Engineering and Management Services Company, Inc. (Contract # 68- 03-3245). The authors would like to express their appreciation to Richard Harvey of the DHEC who provided technical assistance throughout the project. Commercial products are described to support the technical discussion. Their mention does not represent an endorsement by any of the agencies listed.

REFERENCES

Barko, J.W., M.S. Adams, and N.L. Clesceri, 1986. Environmental Factors and Their Consideration in the Management of Submersed Aquatic Vegetation: A Review, *Journal of Aquatic Plant Management.* 24(1):1–10.

Berry J.K., and K.L. Reed, 1987. Computer-assisted Map Analysis: A Set of Primitive Operators for a Flexible Approach, *Technical Papers ASPRS-ACSM Annual Convention,* Baltimore, Maryland, pp. 206–218.

Harvey, R.M., J.R. Pickett, P.G. Mancusi-Ungaro, and G.G. Patterson, 1983. *Aquatic Macrophyte Distribution in Upper Lake Marion.* 1983 Growing Season, Department of Health and Environmental Control, Columbia, S.C., 60 p.

Harvey, R.M., J.R. Pickett, and R.D. Bates, 1987. Environmental Factors Controlling the Growth and Distribution of Submersed Aquatic Macrophytes in Two South Carolina Reservoirs, *Proceedings, Sixth North American Lakes Management Society Meeting,* Portland, Oregon, 32 p.

Jensen, J.R., and B.A. Davis, 1986. Remote Sensing of Aquatic Macrophyte Distribution in Upper Lake Marion, *Technical Papers ACSM-ASPRS Annual Convention,* Washington, D.C., pp. 181–189.

Link, L.E., and K.S. Long, 1978. Large Scale Demonstration of Aquatic Plant Mapping by Remote Sensing, *Proceedings, Twelfth International Symposium on Remote Sensing of Environment,* Vol II, Ann Arbor, Michigan, pp. 907–915.

Martyn, R.D., R.L. Noble, P.W. Bettoli, and R.C. Maggio, 1986. Mapping Aquatic Weeds with Aerial Color Infrared Photography and Evaluating Their Control by Grass Carp, *Journal of Aquatic Plant Management.* 24(1):46–56.

Patterson, G.G., and T.W. Cooney, 1986. Sediment Transport and Deposition in Lakes Marion and Moultrie, South Carolina, *Proceedings, Third International Symposium on River Sedimentation,* The University of Mississippi, pp. 1336–1345.

Radford, A.E., H.E. Ahles, and C.R. Bell, 1983. *Manual of the Vascular Flora of the Carolinas,* University of North Carolina Press, Chapel Hill, N.C., 1183 p.

Seher, J. Scott, and Paul T. Tueller, 1973. Color Aerial Photos for Marshland, *Photogrammetric Engineering.* 39(5):489–499.

South Carolina Aquatic Plant Management Society Newsletter, 1986. Summary of 1986 Aquatic Plant Management Activities in South Carolina. 8(1):4–5.

USGS, 1984. *Bathymetric Map of Lake Marion, South Carolina,* United States Geological Survey, Columbia, South Carolina.

Welch, R., S.S. Fung, and M. Madden Remillard, 1985. *Aquatic Macrophyte Distributions in Lake Marion, South Carolina: 1983– 1984.* Final Report to Lockheed Engineering and Management Services Company, Inc. Contract # 68-03-3245, Laboratory for Remote Sensing and Mapping Science, Department of Geography, University of Georgia, Athens, Georgia, 18 p.

_____, 1986. *Changes in the Distribution of Aquatic Macrophytes: Lake Marion, South Carolina 1972–1984.* Final Report to Lockheed Engineering and Management Services Company, Inc., Contract # 68-03-3245, Laboratory of Remote Sensing and Mapping Science, Department of Geography, University of Georgia, Athens, Georgia, 16 p.

Welch, R., and M.M. Remillard, 1986. *Aquatic Macrophyte Distributions in Lake Marion, South Carolina: June and September, 1985.* Final Report to South Carolina Department of Health and Environmental Control, Contract # EQ-5-427, and United States Environmental Protection Agency, Contract # 5R-1301-NAEX, Laboratory of Remote Sensing and Mapping Science, Department of Geography, University of Georgia, Athens, Georgia, 15 p.

SOLID AND HAZARDOUS WASTE DISPOSAL SITE SELECTION USING DIGITAL GEOGRAPHIC INFORMATION SYSTEM TECHNIQUES

J. R. JENSEN AND E. J. CHRISTENSEN

Geography Dept., Univ. of South Carolina, Columbia, South Carolina 29208 (USA)

ABSTRACT

To identify potential sites for the storage of industrial wastes it is necessary to specify the industrial location constraint criteria, collect the required data, and model this information. This paper first summarizes the type of industrial location constraint criteria that might be imposed. Examples then demonstrate how the required in situ and remotely sensed data can be placed in a geographic information system (GIS) to model and identify candidate waste sites. A case study based on an industrial activity in the southeastern United States is used to demonstrate the utility of using GIS technology.

INTRODUCTION

The storage of solid and hazardous waste can affect air and water quality, land use, and public health. The public is now aware of the quantities of waste materials that must be disposed of and their hazardous nature. It may take years to identify improper disposal practices which affect public health and the environment. Therefore, it is imperative that accurate, thoughtful methods be used to locate industrial waste sites so that their impact on the environment and public is minimized.

Many industries and governments are now faced with the problem of evaluating the impact of existing facilities and identifying economically practical and environmentally safe sites for future waste disposal. In both cases, volumes of multi-disciplinary information must be collected, stored, and analyzed. One approach particularly well suited to the management of such data bases is the Geographic Information System (GIS). A GIS is a digital data base management system designed to accept large volumes of spatially distributed data from a variety of sources (see Fig. 1). The GIS efficiently stores, retrieves, analyzes, and displays the accumulated information according to user-defined specifications.

This paper describes the application of GIS technology to waste disposal facility siting. The general GIS approach discussed can be used to evaluate potential sites and assess impacts from operational or abandoned facilities. The paper focuses on the use of a GIS to structure and integrate a variety of economic and environmental constraints to identify optimum waste site locations. A waste siting situation in the southeastern United States is used to illustrate the process.

© 1986, Elsevier Science Publishers B.V., *The Science of the Total Environment*, v. 56, p. 265–276; reprinted by permission.

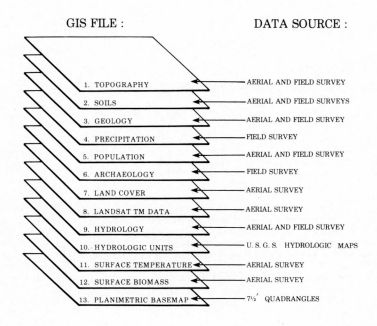

GIS FILE :   DATA SOURCE :

1. TOPOGRAPHY ————————— AERIAL AND FIELD SURVEY
2. SOILS ————————— AERIAL AND FIELD SURVEYS
3. GEOLOGY ————————— AERIAL AND FIELD SURVEY
4. PRECIPITATION ————————— FIELD SURVEY
5. POPULATION ————————— AERIAL AND FIELD SURVEY
6. ARCHAEOLOGY ————————— FIELD SURVEY
7. LAND COVER ————————— AERIAL SURVEY
8. LANDSAT TM DATA ————————— AERIAL SURVEY
9. HYDROLOGY ————————— AERIAL AND FIELD SURVEY
10. HYDROLOGIC UNITS ————————— U. S. G. S. HYDROLOGIC MAPS
11. SURFACE TEMPERATURE ————————— AERIAL SURVEY
12. SURFACE BIOMASS ————————— AERIAL SURVEY
13. PLANIMETRIC BASEMAP ————————— 7½′ QUADRANGLES

Fig. 1. A geographic information system (GIS) consisting of biophysical and cultural files geometrically registered to a planimetric basemap. The information contained in the GIS is derived from a number of sources.

METHODOLOGY

To perform industrial waste site selection using a GIS, the following methodology is suggested: 1) Identify the industrial location constraint criteria (ILCC); 2) Specify the environmental and cultural information required to address the constraint criteria; 3) Collect the required information in analog or digital format using _in situ_ or remote sensing technology; 4) Transfer these point, line, and area data from the original source documents onto a controlled base map having a suitable projection; 5) Digitize the thematic information on the base maps into an acceptable coordinate system, and place the data in a polygon data structure or perform a polygon-to-raster conversion of the data making it suitable for a grid-based analysis; 6) Analyze the thematic information using Geographic Information System (GIS) techniques; 7) Display and statistically analyze geographic areas having a high probability of meeting existing industrial location contraints. Each of these concepts will now be reviewed.

Industrial Location Constraint Criteria

The first step in the GIS site selection process is to specify the industrial location constraint criteria (ILCC). Regulations in the U.S. establish minimum design and siting requirements. Based on these rules, disposal sites are constrained from locating in environmentally sensitive areas including wetlands, 100-year floodplains,

196

permafrost areas, endangered species habitats and recharge zones of sole source aquifers. In fact, about 65 to 75 percent of existing industrial waste disposal sites are in areas particularly susceptible to contamination problems - wetlands, floodplains, and major aquifers (ref. 1).

In this study, potential waste disposal sites were identified using fundamental GIS technology. Site acceptability was judged according to the following criteria:

TABLE 1

Industrial Location Constraint Criteria (ILCC)

The waste site:

- must be located on ground > 100 meters above sea level
- must not be located within 160 meters of a wetland
- must not be located within 160 meters of a sensitive area
- must not be located within 200 meters of an existing production area or waste site
- must be located within 300 meters of a major road.

If a given geographic area met all these criteria, it was judged a suitable candidate for locating the waste site, assuming that no other unknown mitigating factors were present.

There are other biophysical vairables (e.g. soils, geology, depth to groundwater) and cultural criteria (e.g. distance to nearest city having >10,000 people) that could be anlayzed. However, this terse set of industrial location constraint criteria should demonstrate fundamental principles.

Identify the Environmental and Cultural Information Required

Having identified the industrial constraint criteria, it is necessary to specify the types of data that must be incorporated into the analysis. This generally includes both environmental and cultural information on land cover/land use, soil type, elevation, and existing production activites. Table 2 summarizes the environmental and cultural information required in this study.

The wetland information desired are specified according to five U.S. Fish & Wildlife Service non-tidal wetland classes (ref. 2). The sensitive areas include environmental reserves, endangered specie habitats, and archaeological digs. The topographic information is used to stratify the area according to the known 100-year floodplain boundary.

Having identified the constraints and the environmental and cultural information required, it is necessary to collect the data and place it in a format that can be interrogated digitally.

TABLE 2

Environmental and Cultural Information Requirements

Information Required	Code
Existing Production Facilities	
production areas	1
Wetland	
water	2
cypress-tupelo swamp forest	3
emergent marsh	4
scrub-shrub	5
bottomland hardwoods	6
Upland Land Cover	
mixed hardwood and pine	7
young pine	8
old pine	9
cleared fields	10
Transportation Network	
main roads	11
secondary roads	12
railroads	13
utilities	14
Sensitive Areas	
environmental reserves	15
endangered specie habitats	16
archaeological digs	17
Existing Waste Sites	
waste sites	18

Topographic Information
    elevation from a digital terrain model

Collect the Environmental And Cultural Information

This usually necessitates the collection of information using both in situ and remote sensing technology. In this study the geographic coordinates of all existing production areas, waste sites, and sensitive areas were obtained from existing 1:6,000 base maps. Aerial photography (both natural color and color infrared) obtained at a scale of 1:20,000 were interpreted stereoscopically to identify the remaining land cover/land use information, including the wetland data. It is not uncommon to use other types of remotely sensed data to provide the land cover information desired such as that obtained by the Landsat multispectral scanner (MSS) or thematic mapper (TM) sensor system (ref 3.). These data have high geometric fidelity and are easily rectified to standard map projections. However, whenever high spatial detail is required, as is the case with most waste site facility siting, the data will likely be derived from an interpretation of large scale vertical aerial photography (ref. 4).

Occassionally, data from airborne multispectral scanning systems (MSS) are interpreted either visually or digitally to obtain information that will represent a layer in the geographic information system. Unfortunately, such data are usually geometrically distorted due to perturbations in the aircraft's flight altitude, attitude, and other factors. Sometimes the distortions are so great that it is impossible to mosaic the flight lines of the MSS data. Whenever, airborne MSS data are used it is usually necessary to invest considerable resources in image rectification prior to either visual or digital image analysis. Therefore, caution should be exercised when deciding whether or not to incorporate data interpreted from such MSS imagery into a GIS data base.

Transfer Information to Planimetric Base Map

As mentioned, some of the data collected from the various source materials may be geometrically distorted. Therefore, it is imperative that all the environmental and cultural data obtained in an analog map format be transferred to a planimetric base map at a single scale prior to digitization (ref. 5). This is normally accomplished using optical transfer techniques where the point, line, and area polygonal data are transferred from the original scale manuscript documents onto a standard base map in a known and useful map projection (ref. 6). In cases where the data is already in a digital format such as land cover information extracted from digital Landsat data or from digital terrain models (DTMs), the data must be rectified to the base map using coefficients derived from ground control points (GCPs) which are used to warp the digital file into registration with the base map. This is precisely the manner in which the topography > 250' above sea level was extracted from a digital terrain tape of the study area.

In this study, all the analog information were transferred onto U.S. Geological Survey 7.5' topographic maps (1:24,000) because of their planimetric accuracy and because the series is the de facto standard base map in the United States. Furthermore, the Universal Transverse Mercator (UTM) projection in meters northing and easting represents an ideal coordinate system for measuring and comparing the dimension of points, lines and areas.

Digitizaton

There are three fundamental types of geographic data to be stored in a geographic information system (GIS): points, lines, and polygons. These data can be stored in a number of formats, including (1) traditional Cartesian x,y coordinates; (2) in a topological format as node, line segments and polygons (this format is not discussed); or (3) in a grid (raster) format (ref. 6).

The more traditional encoding of points, lines, and polygons is based on the use of Cartesian coordinates such as longitude/latitude founded on the principles of Euclidean geometry. This is often called polygon or vector coding. A diagram showing how a typical map composed of points, lines, and polygons is encoded from its original analog map format into a digital vector format through the process of digitization is found in Figure 2a (ref. 7 and 8). Point, line, or polygon data can also be encoded using a grid-based (often called a raster) cartographic data structure (see Fig. 2b). In a grid-

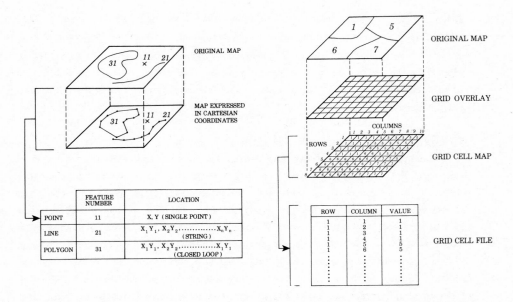

Fig. 2. Diagrams showing (a) the nature of polygon (vector) based geographic information system files in cartesian coordinates, and (b) grid (raster) based geographic information system files stored as arrays.

based data structure, the attribute information is collected within a systematic array of grid cells (ref. 9). These cells are often referred to as picture elements (pixels) and can be analyzed and displayed as if they were 'images', much like a remotely sensed image. Therefore, it is no surprise that grid-based geographic information systems usually have little difficulty incorporating information derived from digital remotely sensed data (ref. 3). Figure 1 illustrates how a number of remote sensing derived files can be registered with other cartographic information in a grid-based GIS format. In order to bring these polygon-based files into registration with the raster remote sensor data, it is necessary to perform a vector-to-raster conversion on the polygon-based files.

In this study, a grid-based (raster) cartographic data structure was used. This necessitated the digitization of every point, line, and polygonal feature on the original 1:24,000 scale maps using a coordinate digitizer that could measure to within ±.001". The polygonal data were then transformed into a grid-based data structure using a polygon-to-raster conversion program with each pixel representing 20 x 20 meters on the ground. Each cell in the grid-based data set is addressable by its UTM x,y coordinate pair.

Analyze the Data Using Geographic Information System Techniques

To extract meaningful information from a GIS data base for industrial location applications, one must be able to query it and ask logical questions, a process commonly referred to as data manipulation. An extensive review of the editing and data manipulation alternatives available on sophisticated GIS systems is found in Reference 7. Here, we will only touch on the most fundamental aspects of data manipulation and provide examples where possible of the methods used in this research.

Frist, it is assumed that the GIS has the capability to make scale and projection changes, remove distortion, and perform coordinate rotation and translation if necessary. The analyst must also be able to browse or roam within the digital data base, selectively windowing in on a region of interest. Then, within this window it should be possible to define an even more specific sub-window within which more detailed questions can be asked. Ideally, this takes place in an interactive, real-time environment with the analyst viewing the data on a high-resolution color monitor.

Once the area of interest is identified, various types of GIS analyses may be performed. Some of the most important include map overlay, map dissolve, polygon overlay for area calculation, and proximity searches. These analysis procedures will be reviewed using both polygon-based and grid-based GIS logic and illustrative examples provided where possible.

Polygon overlay (integration) and dissolve (desintegration) techniques involve the compositing of multiple map files to create a new data set (see Fig. 3). The resultant GIS file contains new polygons formed from the intersection of the boundaries of the two or more sets of original polygon layers. In addition to creating new polygons based on overlay of the multiple layers, it is also possible to calculate the area of specific polygons in relation to their intersection with other polygons. For example, as shown in Figure 3 it is possible to identify the acreage of classes 3, 5, and 9 found only within hydrologic unit #207 using this procedure.

When working with grid-based GIS systems, this same type of analysis is performed using Boolean logic map algebra overlay techniques. Boolean logic uses the operands 'and', 'or', 'not', etc. For example, Figure 4 depicts three files (vegetation, soils, and geology) each additively combined using simple map algebra ('and' logic) to produce a composite map where the higher the value the more suitable the site. Although in this example each variable is equally weighted, it is possible to unequally weight certain variables based on a priori knowledge, increasing or decreasing the relative importance of the variable in the final suitability calculation.

Many times it is necessary to query a GIS file to determine if a given point, line, or area lies within a certain distance of another point, line, or polygon. The most appropriate proximity searches are based on true circular search criteria as shown in Figure 4. Proximity search information is indispensible in most industrial location siting problems where information about adjacency to roads, wetland, sensitive areas etc. is critical to the successful placement of the new facility.

IMPLEMENTATION

For the purposes of this study a grid-based GIS was used to perform the map overlay

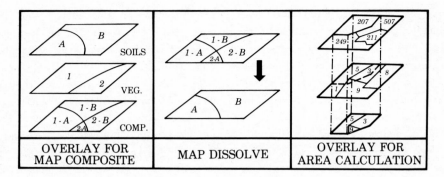

| OVERLAY FOR MAP COMPOSITE | MAP DISSOLVE | OVERLAY FOR AREA CALCULATION |

Fig. 3. Diagrammatic examples of map overlay, map dissolve, and area calculation logic using a polygon (vector) based geographic information system.

and proximity search analyses. The grid cell technique for map manipulation is typically much more efficient both in data storage and in the operation of analytic tasks (ref. 6 and 7).

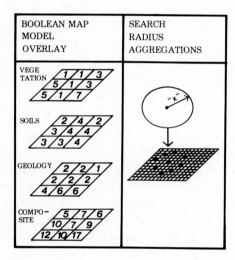

Fig. 4. Examples of (a) Boolean map overlay logic and (b) circular search radius logic.

In this study, grid-cell overlay techniques were used to create a suitability map from an analysis of five (5) input GIS files:

File #1: Land Cover -- (Fig. 5a) Original Land Use/Land Cover File composed of codes 1 - 18. (See Table 2 for codes)
File #2: Topography -- (Fig. 5b) All areas < 100 meters ASL recoded to a value of 20 and those > 100 meters ASL recoded to a value of 0;

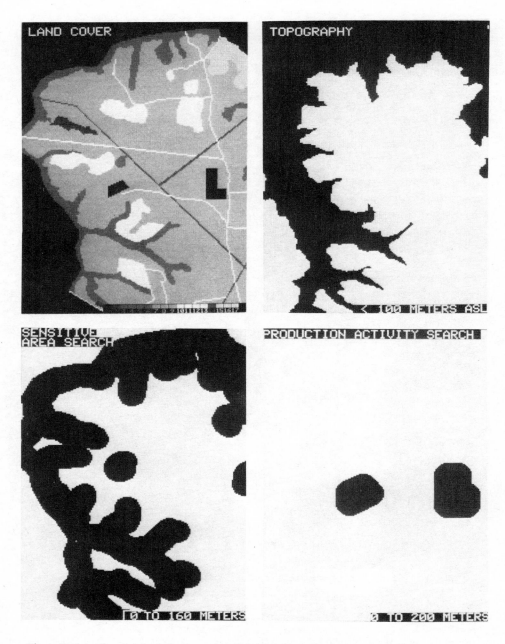

Fig. 5. (a) The original land cover digitized into 18 classes. See Table 2 for the land cover code description. (b) The terrain < 100 meters above sea level is shown in black. (c) All areas within 160 meters of a wetland or sensitive area are shown in black. (d) All areas within 200 meters of an existing production site or hazardous waste site are shown in black.

File #3: Sensitive Areas -- (Fig. 5c) All areas < 160 meters from any wetland or senstive area recoded to a value of 30 and those areas > 160 meters from wetland and sensitive areas recoded to a value of 0;

File #4: Production Activity Search -- (Fig. 5d) All areas < 200 meters from the nearest production area or waste site recoded to a value of 40 and those > 200 meters from the nearest production area or waste site recoded to a value of 0.

File #5: Transportation Search -- (Fig. 6a) All areas < 300 meters from a major road (code #11) recoded to a value of 0 and those > 300 meters to a value of 50.

This necessiated that each of these individual land cover types described above (elevation, wetland/sensitive area, production related areas, and road network) be extracted from the original land cover file and a SEARCH be performed using the specified radius.

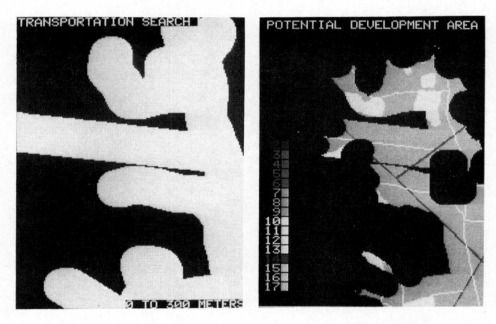

Fig. 6. (a) All terrain > 300 meters from a major road is depicted in black. (b) The potential development area after all industrial location constraint criteria have been evaluated using Boolean map overlay logic. The land cover classes shown are the same as those depicted in Figure 5a. An industrial location analyst must now pick candidate sites within this potential area for more intensive evaluation.

A map algebra program using 'and' Boolean logic was then used to identify only those pixels in the scene that met the desired criteria (see Fig. 6b). The logic is similar to that shown in Figure 4. Those areas within the data set that met the industrial location constraint criteria are displayed in their true land use code (1 - 18) and represent potential development areas which should be evaluated further. Any pixel in the final map with a value less than 0 and greater than 18 did not meet the constraint criteria. Those areas were collectively assigned the color black, effectively masking out the unsuitable areas. Another useful graphic (not shown) identifies the degree of pixel unsuitability by displaying only those areas which did not meet the constraint criteria. In this case the brighter the pixel, the more unsuitable the site.

With the potential development area identified, an industrial site location analyst could identify 50 x 50, 200 x 200 meter etc. candidate facility locations within the proposed burial ground site. The acreage of each land cover type that would be removed if an individual candidate site were selected could also be computed.

SUMMARY

This paper identified the fundamental logic and procedures applied when using a digital geographic information system (GIS) to locate potential sites for the storage of solid or hazardous waste. The importance of the industrial location constraint criteria was addressed first because the user-specified constraints dictate what type of questions must be addressed and what type of data must be collected. With the data requirements specified, it was shown that the data could be collected using in situ or remote sensing methods and then placed in either a polygon-based or raster-based GIS. Boolean logic map algebra and geographic search methods were then applied to a grid-based GIS of an area in the southeastern United States to demonstrate the utility of GIS technology. It was shown that the constraints identified in the industrial location criteria could be effectively evaluated if the proper information were placed in a GIS and appropriate analysis procedures applied to locate candidate hazardous waste sites.

U.S. solid and hazardous waste disposal site regulations require a thorough evaluation of potential impacts. Even though subsurface disposal (once the most common option) is currently discouraged for hazardous waste, land disposal will continue to be the dominant disposal approach for solid and low-level radioactive waste. In many cases, Geographic Information Systems are particularly well suited to managing and analyzing the volumes of diverse multi-disciplinary data needed to make thoughtful waste disposal decisions.

REFERENCES

1. U.S. Department of Energy. Energy and Solid/Hazardous Waste. DOE/EV/10154-2, Washington: 1981.
2. Cowardin, L. M., V. Carter, F. C. Golet and E. T. LaRoe. Classification of Wetlands an Deepwater Habitats of the United States, FWS/OBS-79/31, Washington, 1979, 103 pp.
3. Jensen, J. R. The American Cartographer, 11 (1984) 89 - 100.
4. Jensen, J. R. Introductory Digital Image Processing: A Remote Sensing Perspective, Prentice-Hall, Inglewood Cliffs, N. J., 1986, 450 pp.
5. Avery, T. E. and G. L. Berlin. Interpretation of Aerial Photographs, Burgess Publishing, Minneapolis, Minnesota, 1985, 235-250.
6. Marble, D. F., D. J. Peuquet, A. R. Boyle, N. Bryant, H. W. Calkins, T. Johnson and A. Zobrist. in R. N. Colwell (Ed.), The Manual of Remote Sensing, American Society of Photogrammetry, Falls Church, Virginia, 1983, 923-958.
7. Dangermond, J. in D. Peuquet and J. O'Callaghan (Eds.), Design and Implementation of Computer-Based Geographic Information Systems, International Geographical Union Commission on Geographic Data Sensing and Processing: New York, 1983, 200-210.
8. Marble, D. F., H. W. Calkins, and D. J. Peuquet. Basic Readings in Geographic Information Systems, SPAD Systems Ltd., New York, 1984, pp. 2-57 to 2-78.

Knowledge-Based GIS Techniques Applied to Geological Engineering

E. Lynn Usery* and *Phyllis Altheide*
U. S. Geological Survey, 1400 Independence Road, Rolla, MO 65401
Robin R. P. Deister and *David J. Barr*
Department of Geological Engineering, University of Missouri-Rolla, Rolla, MO 65401

ABSTRACT: Expert system techniques are being investigated to implement a set of rules for geological engineering map production that can handle a variety of input data sources and output classification schemes. The thesis is that a few common data sources such as bedrock geology, agricultural soils, and topography provide the essential data to generate a diverse set of geological engineering maps with a variety of classification schemes. A knowledge-based geographic information system (KBGIS) approach which requires development of a rule base for both GIS processing and for the geological engineering application has been implemented. The rule bases are implemented in the Goldworks expert system development shell interfaced to the Earth Resources Data Analysis System (ERDAS) raster-based GIS for input and output. GIS analysis procedures including recoding, intersection, and union are controlled by the rule base, and the geological engineering map product is generated by the expert system. The KBGIS has been used to generate a geological engineering map of Creve Coeur, Missouri. The computer-generated map compares favorably with a manually produced geological engineering map of the same area and indicates significant promise for KBGIS techniques.

INTRODUCTION

GEOGRAPHIC DATA BASES consisting of remotely sensed digital imagery, digital elevation models (DEMs), and digitized cartographic products can be created interactively with excellent spatial registration and resolution. Geographic information systems (GISs) currently are capable of manipulating such databases to aid in site location, environmental planning, resource management, and other types of decision-making (Tomlinson, 1987). These applications require that large multilayered, heterogeneous, spatially-indexed databases be queried about existence, location, and properties of a wide range of spatial objects (Peuquet, 1984; Smith et al., 1987). While GIS software packages allow these databases to be integrated and manipulated through polygon overlay and other procedures, a comprehensive analysis from initial data entry through final map/product generation requires extensive user interaction and sequential processing steps.

One potential approach to mitigating the requirement for user interaction and sequential processing is to utilize expert knowledge in GIS processing (McKeown, 1987; Robinson and Frank, 1987; Usery et al., 1988). A rule base can be implemented through an expert system inference engine to generate a knowledge-based geographic information system (KBGIS) (Smith and Pazner, 1984). Such a system can be used to solve the critical resource analysis problems of minimizing information management time and maximizing research and application time (Campbell and Roelofs, 1984). A KBGIS might be used to match cartographic representational requirements against a knowledge base concerning classification, symbol schemas, user visual responses, and domain specific knowledge to select the best and most efficient presentation of geographic phenomena (Smith, 1984; Ripple and Ulshoefer, 1987; Bossler et al., 1988).

A research project has been designed to use a knowledge-based approach to solve the problem of generating geological engineering maps from basic Earth resource data. These maps, which usually show surficial and bedrock geologic patterns classified according to engineering suitability for urban development such as waste disposal and building and road construction, are currently produced by manual methods. While use of remote sensing techniques has improved the map production cycle, a system for production of geological engineering maps and tables still requires significant input and interpretation by a skilled geological engineer.

Numerous properties of Earth materials including attributes of soils, geology, hydrology, and topography are considered by a geological engineer when creating a geological engineering map (Varnes, 1974). These properties are usually acquired by field investigation. The amount of time spent in the field is directly reflected in the amount and quality of data collected. Although such information is useful, it may not be cost effective if similar results can be obtained using more basic, previously compiled information such as digitized soils and DEM data.

One objective of this research is to determine the minimum number of Earth resource data sets needed to make engineering judgments about areas in the Midwest. This objective will be accomplished in conjunction with the major project goal of creating a KBGIS that will allow manipulation of these Earth resource data sets to create a geological engineering thematic map. This paper details the preliminary system design with simulated results and the final developed system with actual results of a geologic engineering map generation process compared to a manually produced map.

APPROACH

To create a KBGIS to support engineering geologic mapping requires a thorough understanding of both the concepts of engineering geology and the map generation process. An approach was developed to analyze the basic logic a geological engineer uses in map generation and to transfer that knowledge to an expert system (Figure 1). This expert system approach is being implemented as a series of steps: (1) determine the basic Earth resource information required for production of geological engineering maps, (2) determine the common elements in classification schemes, (3) develop a set of rules for map production, (4) implement the rules in an expert system that will use

*Now with the Department of Geography, University of Wisconsin-Madison, M383 Science Hall, 550 North Park Street, Madison, WI 53706.

Any use of trade names and trademarks in this publication is for descriptive purposes only and does not constitute endorsement by the U.S. Geological Survey.

0099-1112/88/5411–1623$02.25/0
©1988 American Society for Photogrammetry and Remote Sensing

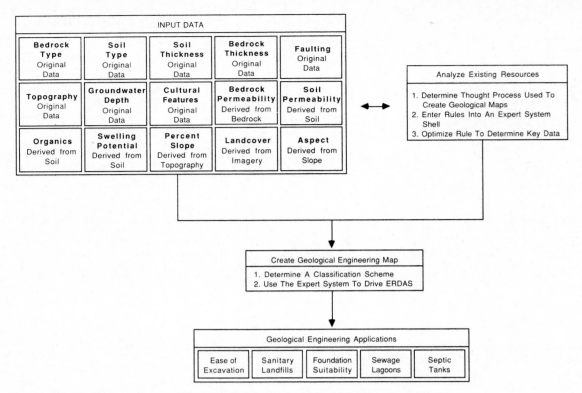

FIG. 1. The approach used to enter the basic logic of geological engineering map production.

the basic Earth resource information to generate various thematic maps, and (5) validate the function of the expert system in a prototype study area.

DEVELOPMENT OF GEOLOGICAL ENGINEERING KNOWLEDGE BASE

To assess the basic information required for the creation of geological engineering themes, the maps in Table 1 were analyzed to determine the logic used in their construction. Information obtained from the different maps and their authors was entered into an expert system shell called 1st Class (Programs-In-Motion, 1987). Data are entered into 1st Class as examples comprised of various soil and rock properties called factors from which a decision tree is built (Figure 2). This type of structure makes 1st Class effective for analyzing different map products for the basic information used in their construction because it allows analysis based on properties instead of the mapped-unit classification.

Preliminary analyses indicate that the basic parameters of agricultural soils and topographic information are sufficient to

Topography	Geology	Plasticity Index	Flooding	Karst	Slope			Result
Flat-marsh	?	20	Frequent or Occasional	No	2			Ia
Low-relief	LS-DO	10	Frequent or Occasional	No	2			Ib
Low-relief	LS-DO	15	Rarely or Never	No	2			Ic
Low-relief	LS-DO	20	Rarely or Never	No	4			Id
Low-relief	LS-DO	27	Rarely or Never	No	2			Ie
Rugged	LS-DO	10	Rarely or Never	No	20			IIa
Rolling	LS-DO	40	Rarely or Never	No	9			IIb
Karst	LS-DO	20	Rarely or Never	Yes	25			IIc
Rugged	LS-DO	20	Rarely or Never	No	40			IId
Roll-rugged	LS-DO	18	Rarely or Never	No	20			IIIa
Roll-rugged	LS-DO	22	Rarely or Never	No	20	• • •		IIIb
Roll-rugged	LS-DO	20	Rarely or Never	No	20			IIIc
Rolling	LS-DO	40	Rarely or Never	Yes	9			IVa
Rugged	LS-DO	25	Rarely or Never	Yes	30			IVb
Rugged	LS-DO	24	Rarely or Never	Yes	30			IVc
Rugged	LS-DO	15	Rarely or Never	No	20			V
Roll-rugged	SH-SS-LS-ST	23	Rarely or Never	No	20			VI
Rugged	SS	12	Rarely or Never	No	30			VIII
Roll-rugged	SH	10	Rarely or Never	No	20			Xa
Gentl-Roll	SH	40	Rarely or Never	No	9			Xb
Low-relief	SH	54	Rarely or Never	No	5			Xc

LS=limestone, DO=dolomite, SH=shale, SS=sandstone, ST=siltstone

FIG. 2. Factors affecting engineering geology are entered as examples in the expert system shell 1st Class. Only 6 of 30 factors are shown.

create geological engineering themes used by field mapping experts. The decision tree aspect of 1st Class minimizes the search involved by considering only those factors which are needed to produce distinct results. In Figure 3, the decision tree uses information from soil surveys, including the plasticity index, knowledge of the presence of karst, flooding potential, and topographic information, with the controlling factor being plasticity index.

Two questions must be addressed when determining a classification scheme: (1) the basic properties to be considered and (2) the potential use to be made of the geological engineering map. It is desirable to have a classification scheme that will provide a standardized system which not only will show distinction between different units of a geological engineering

TABLE 1. GEOLOGICAL ENGINEERING MAPS EXAMINED.

Map	Source
Engineering Geology of the Creve Coeur Quadrangle, Missouri	Rockaway and Lutzen, 1970
Map of Hillside Materials and Their Engineering Character of San Mateo County, California	Wentworth et al., 1985
Engineering Geology of the Northeast Corridor, Washington, D.C., to Boston, Massachusetts	USGS, 1967
Engineering Geology of the Maxville Quadrangle, Jefferson and St. Louis Counties, Missouri	Lutzen, 1968
Engineering Geology of the St. Louis County Quadrangle, Missouri	Lutzen and Rockaway, 1971

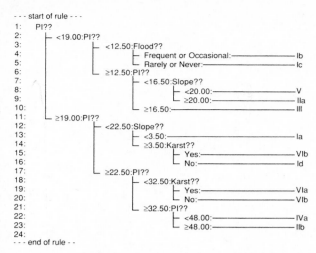

```
- - - start of rule - - -
 1:    PI??
 2:      └ <19.00:PI??
 3:              └ <12.50:Flood??
 4:                      ├ Frequent or Occasional:───────── Ib
 5:                      └ Rarely or Never:───────────────── Ic
 6:              └ ≥12.50:PI??
 7:                      └ <16.50:Slope??
 8:                              ├ <20.00:───────────────── V
 9:                              └ ≥20.00:───────────────── IIa
10:                      └ ≥16.50:───────────────────────── III
11:      └ ≥19.00:PI??
12:              └ <22.50:Slope??
13:                      ├ <3.50:───────────────────────── Ia
14:                      └ ≥3.50:Karst??
15:                              ├ Yes:──────────────────── VIb
16:                              └ No:───────────────────── Id
17:              └ ≥22.50:PI??
18:                      └ <32.50:Karst??
19:                              ├ Yes:──────────────────── VIa
20:                              └ No:───────────────────── VIb
21:                      └ ≥32.50:PI??
22:                              ├ <48.00:───────────────── IVa
23:                              └ ≥48.00:───────────────── IIb
24:
- - - end of rule - -
```

FIG. 3. A part of the decision tree created from the factors in Figure 2.

```
IF Flooding is Frequent or Occasional and
   Slope is <5% and
   Plasticity Index is 10 to 20 and
   Karst is No
THEN  Class is Ia

IF Flooding is Frequent or Occasional and
   Slope is <5% and
   Plasticity Index is <10 and
   Karst is No
THEN  Class is Ib

IF Flooding is Rarely or Never and
   Slope is <5% and
   Plasticity Index is 10 to 20 and
   Karst is No
THEN  Class is Id

         •

         •

         •

IF Plasticity Index is > 40
THEN  Class is Xc
```

FIG. 4. An example of the if-then rules used in the final system design of the KBGIS.

map but also will produce a map accepted and utilized by the geological engineering community. However, it is difficult to devise a single scheme that will suffice for all conditions.

The most favorable approach is to determine a consistent format using the basic Earth resource information that can accommodate several different regions. Consistency is a problem because numerous geological engineering maps cannot be related because the classification schemes vary with respect to the types of factors considered and to the ranges used to define the factors.

To form a basis for classification, several schemes were reviewed and critiqued. Most include bedrock geology and topography information as important factors. However, the factors deviate at that point, with some considering alluvial soils and others considering faults and fractures. In all cases, the products were the result of a combination of factors, and the scheme used by Lutzen and Rockaway (1971) forms a basic structure for a general classification in the Midwest.

The second question to be addressed is the type of geological engineering map to be created. Should the final product be just the factors needed to create a unique entity, or should it make judgments on how the units will function under various conditions? The answer depends on who will use the information. The product is intended for geological engineers to obtain reconnaissance-level information on the geological engineering conditions in an area. The logic used by Lutzen and Rockaway provides an excellent method to meet the needs of the geological engineering community. The classification scheme determines unique entities with respect to their geological engineering contribution, and the entities are evaluated, in a table format, on how they will perform under different site-utilization situations.

Once the basic information was determined and a classification scheme devised, the production rules were entered into an expert system shell. The decision tree results produced in 1st Class were converted to if-then structured rules (Figure 4). The rules from the decision tree were expanded to incorporate the terminology anticipated in the digitized information.

KBGIS DESIGN AND IMPLEMENTATION

In the initial development a rule base was implemented in LISP for producing geological engineering maps (Usery et al., 1988). All GIS processing was performed using the Earth Resources Data Analysis System (ERDAS), a raster-based GIS and image processing system (ERDAS, 1987). The commands to drive ERDAS were generated by the LISP application program from the rule base and the specific pixel values in the Earth resource data files. Using the experience garnered from this initial implementation, a second approach was developed in which a rule base for the geological engineering application and a rule base for GIS processing were implemented in a KBGIS. The KBGIS is frame-based with an inference engine and direct interface to the LISP language as provided by the Goldworks expert system development shell (Figure 5) (Gold Hill, 1987).

In the KBGIS, a frame is used to represent a group of entities with attendant facts. The set of facts or attributes are called the slots of a frame. The actual occurrence of a frame is called an instance. For example, a frame called **Pixel** contains slots for each type of GIS file used in the geological engineering application (Figure 6). An instance of **Pixel** is an actual occurrence with slot values filled.

A rule is knowledge that is used to deduct new facts from existing facts. A fact is referred to as passive knowledge, whereas a rule is referred to as active knowledge. Both fact bases and rule bases are important parts of a knowledge-based system. The mechanism that uses rules and facts to derive new facts is called an inference engine.

In the final system design, ERDAS is used to prepare the Earth resource data files needed to create the geological engineering map. Within ERDAS GIS files, each pixel contains a numeric value between 0 and 255. The expert system uses conceptual values rather than numeric pixel representations.

Conceptual values are actual symbolic language values such as low, medium, and high for slope. The KBGIS uses conceptual values to relate actual GIS overlay pixel numbers to the knowledge base generated from the rule and fact bases. The GIS files, along with a set of rules and a mapping of pixel values to conceptual values, are used by the KBGIS to produce the map. ERDAS is used to display the final map and perform other GIS processing.

Currently, the only GIS operations implemented in the KBGIS are those needed to produce a geological engineering map which are the ERDAS recode and matrix functions. A recode operation in a traditional GIS essentially reassigns classification values to new zones. For example, slope may be represented in 5 percent increments from 0 to 100 percent in a GIS overlay by numbers from 1 to 20. A recoding of the slopes to a total of three classes, 1 for numbers 0 to 5, 2 for numbers 6 to 12, and 3 for numbers 13 to 20, can be performed if the analysis requires values of only low, medium, and high.

In contrast to traditional GIS, the KBGIS recode operation must

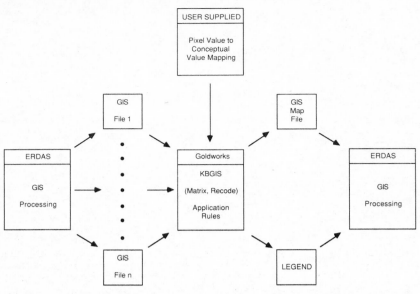

FIG. 5. The system configuration used to generate geological engineering maps from basic earth resource information.

PIXEL
Soils
Flooding
Slope
Plasticity
Karst
GEM

a) A Frame

PIXEL P0512	
Soils	
Flooding	frequent
Slope	<5%
Plasticity	<10
Karst	no
GEM	

b) An Instance

FIG. 6. An example of a frame and an instance of that frame used in the KBGIS. GEM represents the geological engineering map slot.

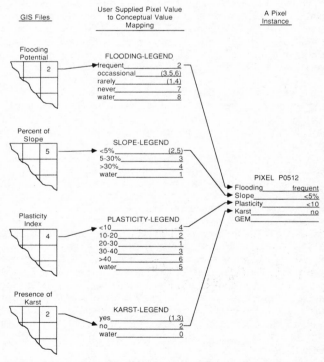

FIG. 7. Frame-based implementation of a GIS recode operation.

handle pixel value to conceptual value and conceptual value to pixel value recodes, as well as the typical pixel value to pixel value recodes. In the slope example above, pixel values 0 to 5 are recoded to conceptual value low, 6 to 12 to medium, and 13 to 20 to high. Also, in a traditional GIS, a new file must be created containing the recoded pixel values, whether it is a final product or an intermediate result. In the KBGIS, a recode only produces files that are final results.

The recode operation in the KBGIS is implemented using frames (Figure 7). A frame is defined for each type of GIS data file such as flooding. The set of conceptual values for the GIS data file are the slots of the frame. An instance of each frame is created in which the slot values are the pixel values to be recoded to the conceptual values represented by the slot names. A slot may have multiple values to allow multiple pixel values to be recoded to a single conceptual value. For GIS files used as input, the user must specify the slot values. For any created GIS file, the slot values are supplied by the system.

To recode a pixel value from a GIS file to a conceptual value, the name from the GIS file instance of the slot containing the pixel value is used. To recode a conceptual value to a pixel value, a pixel value is selected from the slot of the same name as the conceptual value. To perform a pixel value to pixel value recode involves first recoding a pixel value to a conceptual value and then recoding the conceptual value back to a pixel value.

The matrix function in a traditional GIS analyzes two overlays and produces a new overlay containing class values that are coded to indicate how the class values from the original files

coincide or overlap. For example, a 4 by 3 matrix to determine intersection of soil type and vegetative cover would be formed as follows. Class values of 1, 2, and 3 for soil type are assigned to the columns of the matrix. Class values of 1, 2, 3, and 4 for vegetative cover are assigned to the rows of the matrix. Matrix positions are sequentially numbered and become the class values in the output file. Thus, a value of eight in the output file indicates a soil type of two and a vegetative cover of three. This operation allows creation of logical combinations of classes such as union, intersection, complement, or any combination. To matrix more than two files requires a complex series of matrix and recode operations to achieve a result.

In the KBGIS, a matrix operation is required that can operate on any number of files simultaneously. The implementation in the expert system requires creating a frame to represent a pixel (Figure 8). Slot names in the frame represent Earth resource data files from which pixel values will be retrieved and data files to which resultant pixel values will be written. The slot values of pixel instances will be conceptual values. These recoded values correspond to the pixel values as retrieved from the file associated with each slot name.

Rules specific to the geological engineering application are established as if-then constructs. The rules inspect conceptual values in the slots of a pixel instance and set the geological engineering map slot of the pixel instance to the appropriate conceptual value. In this manner, any number of files represented as slots in pixel instances can be matrixed simultaneously to produce a single new overlay. The matrix operation is flexible because not all slots, hence files, need to be inspected by every rule to set the geological engineering map slot value.

Both the matrix and the recode operations, which comprise the GIS processing rule base in the KBGIS, are made more powerful by using the frame data structure and if-then rule constructs. From a user viewpoint, simultaneous processing of multiple overlays and automatic generation of intermediate products greatly simplify the GIS approach to problem solving.

The current implementation of the geological engineering application allows a geological engineering map to be produced from a single overlay with multiple factors or from multiple overlays. For example, separate overlays for flooding, slope, plasticity, and karst can provide input, or a single soils overlay with four factors may be used.

To create a geological engineering map from multiple overlays, the user must specify the information relating pixel values to conceptual values for each of the overlays. Using this information, the KBGIS then recodes the pixel values from the overlays to conceptual values and places them in a pixel instance (Figure 7). The application-specific rules are used to perform the matrix operation to set the geological engineering map slot to a conceptual value for each pixel (Figure 8). The conceptual value of each geological engineering map slot of a pixel instance is recoded to create an ERDAS-format GIS file. A legend is placed in a trailer file so the user knows how to interpret the pixel values generated by the KBGIS.

To create a map from soils data which is a composite of four overlays, the user must describe each soil class (pixel value) in

terms of the properties flooding, slope, plasticity, and presence of karst. The matrix operation is then performed on these pixel value descriptions to set the geological engineering map slot. In essence, the application has derived a recode of the soils pixel values to map conceptual values. The map conceptual values are then recoded to map pixel values. The resulting product is a recode of soils pixel values to map pixel values. To produce the final map, the entire soils GIS file is simply recoded and a trailer file is produced.

RESULTS

To test the initial implementation in LISP, a combination of actual and simulated data over an Aspen, Colorado, test site was used to numerically validate the functioning of matrix and recode operations in a LISP environment. Simulated swelling and soils data were combined with actual overlays of topography, bedrock, and permeability. Although ERDAS was used to perform all GIS processing, the sequence of commands was generated by the LISP application and placed in an audit file. ERDAS then executed the commands from the audit file to create the map. The Aspen test site and the rule base provided the first test of the basic concept and yielded a correct result which was validated by a manual tabulation procedure.

The final system design implements GIS processing within the expert system. For this implementation, a new rule base was developed consisting of two parts, a GIS processing rule base and a geological engineering applications rule base.

This implementation was tested on the Creve Coeur, Missouri, U.S. Geological Survey 7.5-minute quadrangle where actual geological engineering maps exist (Rockaway and Lutzen, 1970). Based on the geological engineering applications rule base, the minimum number of Earth resource files consists of one data set obtained from the Soil Conservation Service soil survey which provides information on four geological engineering factors: soils plasticity index, flooding potential, presence of karst, and topographic information (Anon, 1982). A geological engineering map of the Creve Coeur area was generated from the basic input data and the knowledge bases by the KBGIS. The resulting map was reformatted to an ERDAS GIS file and displayed on a graphics system.

Comparison to the actual geological engineering map that has been digitized was performed by visual techniques using side-by-side displays (Figure 9). The correlation of the computer-generated product and the manually-produced map is believed to be good, especially considering the changes in land use over the 10-year elapsed time since the manually-produced map was compiled (Table 2). Discrepancies are largely a function of the method used for determining the characteristics of each unit.

For example, considerable differentiation exists within the lowland area on the computer-generated map which affects the results of classes Ia, Ib, and Ic. The published map grouped the entire lowland area based completely on its topographic orientation without consideration for variation of engineering properties within the unit. Class Id represents the terrace deposits, or upland deposits, which on the computer-generated map combines areas on topographic divides with similar geological engineering properties. The published map considered only those upland areas adjacent to Class Ic. Classes IIa and IIf differ only in terms of their relative plasticity index. The plasticity index of the soils in the published map were averaged from any available source or location, resulting in an unrepresentative value for the unit. The computer-generated map uses values to a depth of 1 foot for its determination, resulting in a more consistent value for the unit. The published map considered only those areas with visible karst features to be included in class IIc. The computer generated map includes areas adjacent to the visible karst features to be included in class IIc. The difference in the reporting of Class IId is a result of the changing

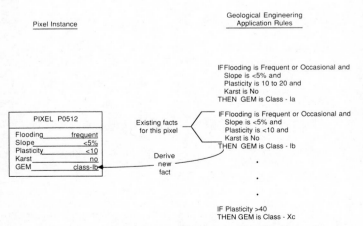

FIG. 8. Frame-based implementation of a matrix operation to set the geological engineering map slot to the correct value according to the knowledge base.

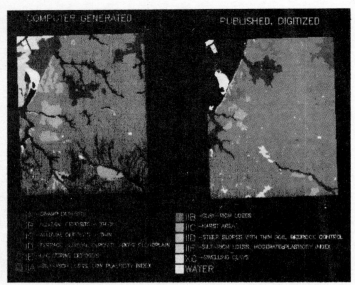

FIG. 9. Comparison of KBGIS generated geological engineering map for Creve Coeur, Missouri, to the manually produced map.

TABLE 2. COMPARISON OF COMPUTER GENERATED AND PUBLISHED GEOLOGICAL ENGINEERING MAPS OF CREVE COEUR, MISSOURI

| | Percent of Total Area | |
Class	Computer Generated	Published Map
Ia, Ib, Ic	7.32	11.32
Id	10.37	2.67
Ie	3.62	2.02
IIa/IIf	2.30	1.95
IIb	67.00	76.89
IIc	5.63	2.11
IId	1.25	0.39
Xc	1.05	0.84
Water	1.47	1.79

land use and development over the 10-year period as well as a more rigid adherence to the classification rules.

CONCLUSIONS

A number of conclusions may be drawn from this work. However, because of the limited testing to date, they must be considered preliminary. It is clear that geological engineering maps can be generated through KBGIS techniques. It appears that a minimum number of Earth resource data sets and a rule base are adequate to generate a map closely resembling a manually produced product. In the case of the Creve Coeur quadrangle, the minimum number is four. Through the course of this work it has become evident that a KBGIS must consist of two knowledge bases, one for GIS processing knowledge and one for application-specific knowledge. The development of additional applications requires implementing a new applications rule base from which the KBGIS can generate a new knowledge base and a new type of product. Future work will consist of using these knowledge bases on additional test sites and expanding the GIS processing knowledge base to implement new GIS operations and applications.

REFERENCES

Anon, A., 1982. Soil Survey of St. Louis County and St. Louis City, Missouri, United States Department of Agriculture, Soil Conservation Service, 137 p.

Bossler, J. D., D. L. Pendleton, G. F. Swetnam, R. L. Vitalo, C. R. Schwarz, S. Alper, and H. P. Danley, 1988. Knowledge-Based Cartography: The NOS Experience, The American Cartographer, Vol. 15, No. 2, pp. 149–161.

Campbell, W. J., and L. H. Roelofs, 1984. Artificial Intelligence Applications for the Remote Sensing and Earth Science Community, Proceedings, Ninth Pecora Symposium on Spatial Information Technologies for Remote Sensing Today and Tomorrow, Sioux Falls, South Dakota.

ERDAS, Inc., 1987. ERDAS-PC and PC Kit: Image Processing System User's Guide, ERDAS, Inc., 430 Tenth St., N.W., Atlanta, Georgia.

Gold Hill, 1987. Goldworks Expert System User's Guide, Gold Hill Computers, Inc., 26 Landsdowne Street, Cambridge, Massachusetts, 282 p.

Lutzen, E. E., 1968. Engineering Geology of the Maxville Quadrangle, Jefferson and St. Louis Counties, Missouri, Engineering Geology Series No. 1, Missouri Geological Survey and Water Resources, Rolla, Missouri, 4 p.

Lutzen, E. E., and J. D. Rockaway, Jr., 1971. Engineering Geology of St. Louis County, Missouri, Engineering Geology Series No. 4, Missouri Geological Survey and Water Resources, Rolla, Missouri, 23 p.

McKeown, D. M. Jr., 1987. The Role of Artificial Intelligence in the Integration of Remotely Sensed Data with Geographic Information Systems, IEEE Transactions on Geoscience and Remote Sensing, Vol. GE-25, No. 3, pp. 330-348.

Peuquet, D. J., 1984. A Conceptual Framework and Comparison of Spatial Data Models, Cartographica, Vol. 21, No. 4, pp. 66–113.

Programs-In-Motion, 1987. 1st Class User's Guide, Programs-In-Motion, Inc., 10 Sycamore Road, Wayland, Massachusetts.

Ripple, W. J., and V. S. Ulshoefer, 1987. Expert Systems and Spatial Data Models for Efficient Geographic Data Handling, Photogrammetric Engineering and Remote Sensing, Vol. 53, No. 10, pp. 1431–1433.

Robinson, V. B., and A. U. Frank, 1987. Expert Systems for Geographic Information Systems, Photogrammetric Engineering and Remote Sensing, Vol. 53, No. 10, pp. 1435–1441.

Rockaway, J. D., Jr., and E. E. Lutzen, 1970. Engineering Geology of the Creve Coeur Quadrangle, St. Louis County, Missouri, Engineering Geology Series No. 2, Missouri Geological Survey and Water Resources, Rolla, Missouri, 19 p.

Smith, T. R., 1984. Artificial Intelligence and its Applicability to Geographical Problem Solving, The Professional Geographer, Vol. 36, No. 2, pp. 147–158.

Smith, T. R., and M. Pazner, 1984. Knowledge-Based Control of Search and Learning in a Large-Scale GIS, Proceedings, International Symposium on Spatial Data Handling, Zurich, Switzerland, pp. 498–519.

Smith T., D. Peuquet, S. Menon, and P. Agarwal, 1987. KBGIS II: A Knowledge-Based Geographical Information System, International Journal of Geographical Systems, Vol. 1, No. 2, pp. 149–172.

Tomlinson, R. F., 1987. Current and Potential Uses of Geographical Information Systems. The North American Experience, International Journal of Geographical Information Systems, Vol. 1, No. 3, pp. 203–218.

U.S. Geological Survey, 1967. Engineering Geology of the Northeast Corridor Washington, D.C., to Boston, Massachusetts: Bedrock Geology, U.S. Geological Survey Miscellaneous Investigations Map, I-514-A.

Usery, E. L., R. R. P. Deister, and D. J. Barr, 1988. A Geological Engineering Application of a Knowledge-Based Geographic Information System, Proceedings, American Congress on Surveying and Mapping/American Society for Photogrammetry and Remote Sensing, Vol. 2, Cartography, St. Louis, Missouri, pp. 176–185.

Varnes, D. J., 1974. The Logic of Geologic Maps with Reference to Their Interpretation and Use for Engineering Purposes, U.S. Geological Survey Professional Paper 837, 47 p.

Wentworth, C. M., S. Ellen, V. A. Frizzell, Jr., and J. Schlocker, 1985. Map of Hillside Materials and Description of their Engineering Character, San Mateo County, California, U.S. Geological Survey Miscellaneous Investigations Map, I–1157–D.

SECTION 7
GIS Selection

Overview

It is very important to provide information and guidelines to help people involved with the implementation and procurement of GIS make the best decisions possible. The article by Stefanovic and Drummond gives a step-by-step approach for evaluating the performance of a GIS. Their selection procedure matches the specifications of the GIS with the perceived needs of the user.

Suggested Additional Reading

Goddchild, M.F. and B.R. Rizzo 1987. Performance Evaluation and Workload Estimation for Geographic Information Systems. *International Journal of Geographical Information Systems*. 1(1):67–76.

Guptill, S.C. (ed.) 1988. *A Process for Evaluating Geographic Information Systems*. U.S. Geological Survey Open-File Report 88–105, 57p.

Selection and evaluation of computer-assisted mapping and geo information systems

Pavao Stefanovič* and Jane Drummond*

ABSTRACT

This article describes, in steps, the procedures to be followed when selecting and evaluating the performance of a computerized mapping system, prior to its purchase. It is recommended that selection be based on the ability of the system to meet specifications derived from the production and system requirements matching perceived goals. Because the purchaser of digital equipment so often does not get what he expects, however, those stages in the procedure where the purchaser must exercise control are indicated. Although long recognized as the point at which the purchaser can claim control, benchmarking is questioned and its replacement by performance evaluation based on published calibration test results is advocated.

Purchasing a computer-assisted cartography system or a geographic information system can be very expensive, and may leave those responsible for the purchase open to criticism from their less involved colleagues. For very large production systems to be established in national mapping organizations, careful benchmarking, acceptance testing, and post implementation reporting may be an official requirement. Benchmarking, and its mirror-image, acceptance testing, however, are expensive processes, and may not be the most effective way of establishing a system's suitability—especially when funds are limited. Intelligent reading of computing literature and company literature may provide more information than most benchmarks. A systematic selection procedure incorporating some type of performance test, and a carefully prepared contract should be sufficient to reassure the customer that he is getting what he expects. A benchmark which attempts to mimic the production environment in an applications-oriented test is unlikely to truly do so, and will lead to misplaced assurance. It is the purpose of this article to describe selection procedures and performance evaluation for computer-assisted cartography systems.

THE SELECTION PROCEDURE

SPECIFY THE PRODUCTION AND SYSTEM REQUIREMENTS

First, the overall objectives of the production system to be acquired must be established. These should be thoroughly understood. The system objectives may include one or more of the following:
- to produce specific products
- to acquire a reliable system
- to acquire a cost-effective system
- to improve the existing situation
- to acquire a modular system
- to acquire a system which is expandable
- to acquire a system which can be upgraded
- to acquire "state-of-the-art" hardware
- to acquire a maintainable system
- to acquire a system which is compatible with other systems.

When more than one objective is accepted, priorities must be established.

General production requirements need to be identified at an early stage because they represent the basis for the later development of system specifications. This involves in the first place the establishment of requirements. The present operational and managerial requirements have to be identified and examined. Projection of requirements into the future is also essential. Depending on the complexity of the operational environment, several years will be required to complete the system acquisition and implementation, and new needs may arise in the meantime. The system has to be designed to satisfy the future needs at least for the coming three to eight years, *ie*, to the point where the system is expected to become obsolete.

INVESTIGATE AVAILABLE TECHNOLOGY FOR DATA GATHERING, PROCESSING AND PRESENTATION, USING LITERATURE, CONSULTANTS AND EXPERIENCE IN SIMILAR PROJECTS

Before starting any detailed system specifications, it is necessary to get acquainted with the "state-of-the-art" and the market, to determine what is and may soon be available in the field of interest. Such information is often already available in-house or with consultants, and this may speed up the whole procedure. Also certain journals specialize in providing regular surveys of particular components—such as workstations or graphic screens.

This knowledge is important because there is a high rate of progress in the computer industry and especially in computer graphics.

DESCRIBE DATA, DATA MANIPULATION PROCEDURES AND THE REQUIRED PRODUCTS

Available data or the data to be captured need to be precisely described. A data specification may already exist, but computer processing requires precise and detailed descriptions. The ideal situation occurs when such descriptions result in a data dictionary which later becomes the basic link between an operator

* Department of cartography, ITC

© 1987, International Institute for Aerospace Survey and Earth Sciences, *ITC Journal*, n. 1, p. 39–44; reprinted by permission.

and a data base. Furthermore, the purpose and results of operations must be established, and also the form in which results should be presented (maps, tabular forms, etc). If an exchange of data with other organizations is planned, an exchange format should be defined.

At this point, the workload of the system over a certain period into the future, say eight years, must be estimated. This will be used to define the performance characteristics of the system to be purchased, while the analysis of the production procedures will be used to define functions in the system.

SELECT AND DESCRIBE PRODUCTION PROCEDURES; EVALUATE MANPOWER REQUIREMENTS

At this stage, the production procedures can be planned and described. Actual needs do have to be satisfied but an important goal at this stage is to define the final production system configuration, utilizing to the fullest extent available hardware and software. The production process for each product—either present or planned—must be reduced to its subtask components. Often various procedures share common subtasks. Too fine a subdivision into subtasks should be avoided as this may narrow the available choice.

For every subtask, all functional capabilities have to be identified and described, preferably at the conceptual level. Without a very good reason, it is better not to specify any particular algorithm or hardware components. All that is needed is a clear definition of each subtask in terms of all functions to be performed, links with other relevant subtasks, input and output, interface with the operator, and expected performance.

By this stage, it should be possible to estimate the total workload, manpower, and training requirements for a future system. Manpower requirements especially must be identified as soon as possible. Care should be taken that discussions on the manpower problem include representatives of all staff likely to be involved with the new acquisition, to avoid possible later personnel conflicts. The costs of staff training should be included in the financing plans.

LIST SYSTEM SPECIFICATIONS

All these activities lead to a list of system specifications which will be used in the first contact with the vendors. Such a list should therefore be clear and concise. The vendors must be informed about general system objectives and the function of each subtask, and also the expected system performance and data volumes.

The official status of such specification varies from country to country. In some situations, it is not permissible to change the specifications after the call-to-tender has been issued. Any changes will then require a further call-to-tender. In other circumstances, it may be permissible to alter the specifications as responses are received (this situation is re-

ferred to below). Some users (especially educational users) are interested in procedures, whereas others are interested in products, and this will be reflected in the specifications. There is thus no such thing as an ideal specification structure, but an example of a specification has been provided by the Saskatchewan Department of Tourism and Natural Resources [3].

DEVISE CRITERIA FOR VENDOR SELECTION

Not only the product to be purchased but also the vendor should be evaluated. Thus after the list of specifications has been established, it is necessary to devise criteria for vendor selection and compile a list of all potential vendors. It is important to include all possible vendors even if they do not seem very promising initially. The computer market is dynamic and good new products appear with little warning. The nature of digital equipment is such that it requires the vendor's support throughout its lifetime. Thus each vendor has to be evaluated on the basis of reliability and stability (including financial stability). There is always a risk that support may cease.

The frequency with which a vendor upgrades his products should be evaluated because it shows rate of adaptation to development. On the other hand, a vendor may be providing system "upgrades" inconveniently often.

The selection of a vendor can be done in two steps—preliminary selection and final selection. Preliminary selection will be based on commercial literature obtained from vendors, the expertise of persons frequently involved in evaluation, journal articles and, most useful, discussion with organizations already using similar equipment. The preliminary selection eliminates less serious candidates, is mainly paper work, and it should not be too expensive. It can be based on a scoring technique completed in two stages. At first a global list of all subtasks is used and each subtask is then resolved into a number of components: hardware, software, manpower, reliability, performance, costs, etc. In such a way the whole evaluation may be subdivided among specialists in various subtasks. The range of scores per item must be discussed and decided upon together with the eventual weighting of components.

Very often, especially with complex systems, a single vendor will have no completely satisfactory solution for all subtasks in the list. During the preliminary selection procedure, one important issue is whether a combination of vendors should be considered for short-listing. Compatibility among modules originating from different vendors must be assured. Manufacturers are increasingly achieving compatibility between systems, so that vendor combinations will become more popular. When such a combination is favourable, however, it is advisable (if possible) to sign a contract through a single vendor who carries responsibility for the whole system and who may then subcontract to all other participating vendors.

For final selection, more thorough investigations

are needed, and this is carried out after the call-to-tender and resulting proposals.

ISSUE A CALL-TO-TENDER

All vendors passing the preliminary selection are invited to submit proposals for the required system, answer any specific questions and explain how the stated objectives can be reached with the proposed system. A deadline for the submission of proposals must be specified. It is also useful to inform the vendors about financial limitations. When the organization in question does not have sufficient staff capacity or sufficient expertise to complete the selection procedure, a consulting bureau can be contracted to work in close cooperation with the purchasing organization.

RECEIVE PROPOSALS AND EVALUATE THEM ACCORDING TO DEVISED CRITERIA

Within the prescribed period, the vendors are expected to provide information on the extent to which the proposed system is capable of performing the prescribed work by subtask, price details, warrant conditions, conditions and prices for maintenance, and delivery period.

Some vendors may immediately concede an inability to deliver. For some others, there will be an indication that delivery will merely be difficult. The call-to-tender, however, should produce some potential vendors and the purchaser will have to devise a method for objectively assessing the response and the vendor's ability to provide the required system. To resolve uncertainties, bidders meetings can be organized and system demonstrations may be suggested. It should be remembered that manufacturers frequently have seminars, open days, workshops, etc, and if these are attended, special (and expensive) demonstrations will not be needed. In addition, during the evaluation of proposals new solutions might become obvious or some planned procedures might emerge as either impossible or too expensive. In such instances, it is better to adjust the original specifications to accommodate new situations. A checklist—of all specifications—will reveal those vendors who are able to meet all, or most, requirements. Such vendors can be short-listed. It is unlikely that the final short-list will include more than six vendors.

Before the short-list is completed and used, it must be understood what exactly the short-listed vendors offer, what their shortcomings are, and where further negotiations may be possible. If satisfactory solutions are not found for some subtasks, such subtasks may be contracted out.

DRAW UP A DRAFT CONTRACT; NEGOTIATE WITH THE SHORT-LISTED VENDORS; DESIGN THE BENCHMARK TEST

Computer marketing started in the 1950s as a leasing business. Contracts were thus traditionally provided by the supplier. In the computing business, this tradition has died hard. It is in the purchaser's interest to provide his own contract, however, or an unbiased third party (such as consultants, or a government department specially established to aid in the purchase of high technology) may prepare a model contract. The contract should cover: payments; hardware, software, or both; hardware modifications, software modifications, or both; relocation; transportation; training; maintenance both during and after the warranty period; and user and technical documentation. (A full checklist for both hardware and software contracts is provided at the end of this article.)

Production of the contract may be iterative. This is because the system specifications should be built into the contract and these may be modified as negotiations proceed. Also the size and staggering of payments will change as negotiations proceed. To measure performance requires what used to be called a "benchmark test". Although this has never been a well-defined concept, basically an attempt is made to mimic the production environment and thereby assess a system's capability. The design of benchmark tests involves the selection of test data sets typical of the planned production, selecting operations to be performed and devising the evaluation criteria. The actual execution of benchmark tests will differ from vendor to vendor in fine detail, but efforts should be made to have the benchmark vary as little as possible from one vendor to another. Sometimes detailed standard benchmark tests are prepared by vendors. Then the results of such tests need only be evaluated and made compatible with tests from other competitors. The term benchmark is now falling into disrepute, to be replaced by other terms, such as "calibration tests". These performance tests will be discussed below.

SELECT VENDOR BY DETAILED CRITERIA ASSESSMENT OR COMPARATIVE BENCHMARK (IF SUITABLE RESULTS ARE NOT KNOWN)

Vendor selection is done next. This is a function of system performance and "exposure rating" (ie, the total cost to an organization of implementing a particular system). System performance rating will be looked at below.

Exposure rating is often neglected. It requires systematic effort and can be carried out at the purchaser's site. There is little glamour attached to it. It involves determining the cost to the user of implementing the system. This is not only the purchase cost, but also other costs such as building modifications, staff retraining, staff relocation, production loss during installation, interest rates, etc. The purchaser's exposure costs with time should also be estimated, so that in the event of vendor delay, or withdrawal, expenses can be calculated for inclusion in the contract. These concepts become increasingly important for larger and more complex systems.

NEGOTIATE THE FINAL CONTRACT; DESIGN THE ACCEPTANCE TEST; COMPLETE THE CONTRACT

The first-ranked vendor is approached and told he is first-ranked. At this point, all information required to complete a satisfactory contract should be available and a satisfactory contract negotiated. If a satisfactory contract cannot be negotiated, the second-ranked vendor should be approached. This can be repeated for as long as satisfactory vendors are available—but it is unlikely that there will be many of these. It should be remembered that up to this point *no* investment will have been made in a *particular* vendor's equipment, through either site development or staff training. General preparation or education may, of course, have begun.

A series of acceptance tests are performed on systems. The manufacturer performs some himself at the manufacturing site and at the purchaser's site. Under some circumstances, acceptance testing may end here. Contracts often allow the purchaser to perform his own acceptance testing within the first five or 10 working days after receiving the equipment, however. This is often a repeat of the benchmark test. On completion of the acceptance test (and therefore official acceptance of the system by the purchaser) the warranty period begins. The timing and significance of the various acceptance tests should be built into the contract.

The contract must then, of course, be signed. It then remains to follow out the terms of the contract. In the contract there will be the details of the delivery date(s), acceptance testing procedure(s) and date(s), payment dates, vendor's training date(s), warranty period(s), maintenance, document supplies, software upgrading and hardware upgrading.

ORGANIZE OPERATOR STAFF TRAINING AND STAFF SELECTION

If we consider that staff are found at three levels in the computer environment—managerial, supervisory and operational—then staff members at the first of these levels will have already been involved with the planning and selection of the new system. Some staff who will later function at the supervisory level may also be in place. There is no point, however, in training operator level staff until they are just about ready to start operating. Also it may not be possible to identify suitable operator staff from the existing pool of staff until some operator experience has been gained. Operator staff training can be given by the manufacturer or by supervisors, but should be given on an installed system, on the system which will be used, and to more staff than will eventually become functioning operators.

DELIVERY, PERFORM ACCEPTANCE TEST, COMPLETE PAYMENT

The site should be prepared by the time the system is delivered so that the accompanying engineers are immediately able to install the system and run their own site acceptance tests. Failure to immediately install a purchased system may prevent the application of the acceptance test procedures.

If agreed in the contract, the customer should then run his own acceptance test within the specified number of working days.

Details of payment are, of course, outlined in the contract. Payment will normally be delayed only until after the successful completion of the acceptance test for the very smallest system. Usually, the final 10 or 20 percent of the payment is made after a successful acceptance test, with the initial 80 to 90 percent of the payment having been made at agreed points through the purchase process and leading up to the departure of the system from the vendor's premises. An early payment of 20 to 30 percent may be required as soon as the contract is signed and before the vendor starts to make a specific investment in the customer.

PERFORM OPERATIONAL TEST DURING THE WARRANTY PERIOD

The acceptance test may be performed under some time pressure, and may merely be a repeat of a benchmark test. Unless the benchmark test was extremely well designed, it may not truly reflect all applications or the operating environment. It is therefore necessary to ensure during the warranty period that as many necessary repairs as possible are carried out. To do this, there should be as much production variety as possible, and this may mean carrying out a large number of small and varied projects during this period.

PERFORMANCE EVALUATION

In the past, much time, effort and money was spent benchmarking. There appears now to be a general feeling, especially among smaller manufacturers—but, among others too [2]—that these have been not only spent, but wasted. The purpose of these tests seems to have been for the potential customer to prove to himself, in a way he understands, that the system can do what it is supposed to do. This proving is done in a small applications-oriented project. A benchmark test might thus involve the capture, processing and plotting of cartographic data from a particular map sheet. Each potential vendor would perform this task and the quality of the output and the time taken to complete the task would be used in an attempt to rank the vendors. The test might be performed in the presence of the potential client. With so many variables (operator skill, CPU power, task complexity, system multi-usage, etc), however, it is unlikely that any benchmark will replicate the eventual production circumstances. The scientific and managerial advantages of benchmarking have therefore been questioned. It is likely that the same or better information could have been gathered, more cheaply, by reading published commercial material.

Attempts have been made to make benchmarks more objective. These include reducing the test to very specific tasks—such as the time taken to scan, vectorize, feature code, and correct contours from a production separation—or producing a synthetic benchmark. A synthetic benchmark uses synthetic test data which can give a statistically controlled distribution of map features on the basis of their complexity. Such a synthetic product does not reveal real details of the potential purchaser's working practices. This is important because one of the criticisms levelled at the whole benchmark process is that the information remains confidential. It also has emerged that the confidentiality problem lies not only with the vendors, but also with the potential purchasers who may not wish the nature of their data or their working practices to be revealed. Results from synthetic benchmarks seem more publishable.

An alternative to benchmarking is calibration. Envisaged system usage and workloads can be reduced to several data sets of varying volume and complexity, on which relevant operations are performed. Each operation, and the operating conditions, must be sufficiently well understood that the measured findings for each level of data set can be used to predict performance in the operating environment. It is suggested that the measured findings be CPU time, elapsed time, page fault rate, I/O rate, for the average, minimum, and maximum data sets in terms of volume and complexity, in an environment consisting of 1 to n terminals.

Certain functions are now well established in the geographic information system environment—such as polygon overlay or buffering. Others—such as route finding—continue to emerge. Standardized calibration tests for each function could be performed by a vendor whenever it has a new function. The results could be published. Potential purchasers will then merely have to predict performance in their own environment, and expensive benchmarks will be avoided.

REFERENCES

1 Brandon, D H and S Segelstein. 1976. Data Processing Contracts. Van Nostrand, New York.

2 Deen, S M. 1985. Principles and Practice of Database Systems. MacMillan Publishers Ltd, London.

3 Tomlinson, R F and R Boyle. 1981. The state of development of systems for handling natural resources inventory data. Cartographica, Vol 18, No 4, pp 65-95.

RECOMMENDED READING

Cassidy, B J. 1985. The Rat that Died Twice. Oracle Corporation Europe, Naarden, The Netherlands.

Goodchild, M F and B R Rizzo. 1986. Performance evaluation and workload estimation for geographic information systems. Proc Internat Sem on Spatial Data Handling, Seattle, Washington, USA.

Marble, D F and L Sen. 1986. The development of standardized benchmarks for spatial data systems. Proc Internat Sem on Spatial Data Handling, Seattle, Washington, USA.

RESUME

Cet article décrit les procédures à suivre pour sélectionner et évaluer la performance d'un système de cartographie informatisé avant son achat. Une sélection devrait être basée sur la capacité du système à répondre aux spécifications dérivées de la production et aux exigences du système à satisfaire aux objectifs fixés. C'est parce que, si souvent, l'acheteur d'équipement digital n'obtient pas ce qu'il avait escompté que les étapes du procédé dans lesquelles il doit exercer un contrôle sont indiquées. Bien qu'il soit reconnu depuis longtemps comme le point sur lequel l'acheteur peut réclamer un contrôle, le jeu de tests est remis en question. Son remplacement par une évaluation de performance basée sur des résultats de test de calibrage publiés est préconisé.

RESUMEN

Este articulo describe, en etapas, el procedimiento que debe seguirse cuando se selecciona y se evalua el trabajo de un sistema computerizado para hacer mapas, antes de su compra. La seleccion deberia basarse en la capacidad del sistema para hacer frente a las especificaciones derivadas de la produccion y los requerimientos del sistema para lograr los objetivos deseados. Debido a que el comprador del equipo digital a menudo no obtiene lo que el espera, sin embargo, esas fases en el procedimiento donde el comprador debe ejercer control estan indicadas. A pesar que se reconoce por mucho tiempo como el punto en el cual el comprador puede establecer control, estudios comparativos se ponen en duda se sugiere en su reeplazo por evaluaciones de perfomancia basadas en resultados de pruebas de calibracion publicadas.

Checklist for the contents of a hardware contract (based on [1])

Hardware components

(1) Hardware component specifications
(2) Hardware configuration
(3) System performance
(4) Inclusion of the vendor's original proposal
(5) Inclusion of hardware manuals
(6) Guarantees of compatability
(7) Guarantees of interface
(8) Ability to modify equipment in the field

Service and Supplies

(9) Supply requirements
(10) Supply specification
(11) Acceptable supply vendors
(12) Guaranteed supply availability
(13) Power requirements and specifications
(14) Specifications of the ambient environment
(15) Cabling requirements
(16) Support to be supplied
(17) Staff calibre
(18) Training program
(19) Availability of training
(20) Continued availability of training
(21) Availability of instructors
(22) Availability of educational materials
(23) Rights to future courses
(24) Rights to teach courses internally
(25) Test time availability
(26) Availability of documentation
(27) Rights to future documentation
(28) Rights to reproduce documentation

Costs and Charges

(29) Terms of agreement
(30) Payment
(31) Hold-backs of payment
(32) Tax applicability
(33) Warranty of new equipment
(34) Investment tax credits
(35) Methods of charging and charging types
(36) Credits for malfunctions and other features
(37) Price protection prior to delivery

(38) Price protection during the contract term
(39) Price protection for expansion equipment
(40) Price protection for supplies and services
(41) Price protection for maintenance services

Reliability and Acceptance

(42) Reliability parameters
(43) Guarantees of performance
(44) Warranty
(45) Right to replace components
(46) Backup availability
(47) Disaster availability
(48) Acceptance testing

Maintenance

(49) Available maintenance types
(50) On-site maintenance
(51) Response time for off-site maintenance
(52) Access requirements
(53) Space and facilities
(54) Spare parts
(55) Continuity and maintenance renewal rights
(56) Reconditioning on resale
(57) Rights to training of own maintenance staff
(58) Rights to purchase spare parts
(59) Rebates for failures not corrected promptly
(60) Malfunction and correction reporting
(61) Subcontract restriction
(62) Engineering changes
(63) Rights to schedule preventive maintenance
(64) User right to perform maintenance functions on rental contracts

Delivery

(65) Delivery dates, by component or group
(66) Option for early delivery
(67) Right to delay or postpone delivery
(68) Notice of pending delivery
(69) Irreparable delays
(70) Site preparation responsibility
(71) Installation responsibility
(72) Risk of loss prior to installation
(73) Right to cancel components prior to delivery
(74) Substitution of components prior to delivery

Rights and Options

(75) Unrestricted use and function
(76) Unrestricted location
(77) Right to make changes and attachments
(78) Interface with equipment of other manufacturers
(79) Right to upgrade hardware
(80) Right to upgrade software
(81) Hardware trade-in
(82) Title transfer
(83) Purchase option

(84) Rental credit
(85) Protection of purchase price under purchase offer
(86) Availability of expansion units
(87) Alternative sources of supply

Damages

(88) Damages incurred if contract terminated prior to delivery
(89) Failure to deliver
(90) Liquidation of damages
(91) User costs

Checklist for the contents of a software contract (based on [1])

Performance

(1) Specifications
(2) Definition of documentation
(3) Availability of future documentation
(4) Right to reproduce documentation
(5) Rights to future options
(6) Run time
(7) Facility requirements
(8) Error detection
(9) Upgrades
(10) Source availability and access
(11) Right to modify programs
(12) Compliance with standards

Financial

(13) Charges by type
(14) Payment terms
(15) Terms of licence
(16) Tax status of software
(17) Rights upon business termination
(18) Price protection on licence fee
(19) Price protection on other charges
(20) Availability of financial statements

Installation and support

(21) Delivery
(22) Delivery failure
(23) Installation and modification assistance
(24) Support requirements
(25) Acceptance
(26) Destruction on termination

Warranties

(27) Guarantee of ownership
(28) Copyright, patent, or proprietary right infringement
(29) Guarantee of operation
(30) Free maintenance
(31) Freedom of use

SECTION 8
GIS Literature and Newsletters

Overview

The purpose of this section is to direct the reader to additional sources of written information on GIS. First, Ripple describes sources of GIS literature including symposia proceedings, scientific journals, and books. In the second paper, Merchant and Caron provide an international guide to newsletters found on GIS and related topics.

GEOGRAPHIC INFORMATION SYSTEMS LITERATURE

by

William J. Ripple
Environmental Remote Sensing Applications Laboratory
College of Forestry
Oregon State University, Corvallis, Oregon 97331-5714
1988

Introduction

Information on geographic information systems (GIS) can be found in a wide variety of sources including conference and symposium proceedings, journals, books, and newsletters. The purpose of the following discussion is to provide a brief guide to GIS literature. This section only covers recent sources of literature. For older references on GIS, please see Peucker (1980) and for applications articles and references see Ripple (1987). A two volume consolidated bibliography on land-related information systems was published by the Alberta Bureau of Surveying and Mapping for 1980-1983 (Volume 1) and 1984-1986 (Volume 2). Copies of these bibliographies can be obtained from Maps Alberta, Department of Forestry Lands and Wildlife, 2nd Floor, North Petroleum Plaza, 9945-108 St., Edmonton, Alberta, Canada T6B 2T5.

Conference Proceedings

The number of conferences at which papers are presented on GIS topics is growing rapidly. In recent years, we have seen GIS presentations become a major part of the semi-annual conventions of the American Society for Photogrammetry and Remote Sensing (ASPRS) and the American Congress on Surveying and Mapping (ACSM). The Urban and Regional Information Systems Association (URISA) also publishes a proceedings of its annual meeting. Typically a number of GIS papers are presented at the annual meeting of the Association of American Geographers (AAG). Every two years, there are two international meetings related to GIS. These are the International Symposium on Computer-Assisted Cartography (Auto-Carto) and the International Symposium on Spatial Data Handling. The Auto-Carto meeting is typically held in conjunction with alternating annual meetings of ASPRS/ACSM. Special GIS meetings also occur from time-to-time. For example, a meeting on GIS in government held in 1986, yielded a two volume hard-bound set of proceedings (Opitz 1986). Researchers from both the United States and Australia cooperated on a workshop on the Design and Implementation of Computer-based Geographic Information Systems, the proceedings of which were published by the IGU Commission on Geographic Data Sensing and Processing (Peuquet and O'Callaghan 1983). In 1988, there was a GIS symposium sponsored by the National Academy of Sciences, the U.S. Geological Survey, and the Association of American State Geologists. In 1986, ASPRS sponsored a Geographic Information Systems Workshop in Atlanta, Georgia. In 1987, this was expanded and held in San Francisco as GIS '87. In 1988, the meeting (GIS/LIS '88) was even larger with AAG and URISA also serving as co-sponsors in San Antonio, Texas. These meetings are contributing large sets of proceedings to the GIS literature.

Refereed Journals

The number of professional journals carrying articles on GIS is also increasing rapidly. In 1987, the **International Journal of Geographical Information Systems** was initiated. Numerous fundamental GIS articles can also be found in cartography periodicals. Examples include, The **American Cartographer** (Newcomer and Szajgin 1984; White 1984), **Cartographica** (Blakemore, 1984; Chrisman 1984a; Mark, 1984; Peuquet, 1984b) and **the Cartographic Journal** (Wilkinson and Fisher 1987). A number of GIS articles can also be found in remote sensing journals, such as **Photogrammetric Engineering and Remote Sensing**, which published a special GIS issue in October of 1987 and again in Nobember of 1988, and the **International Journal of Remote Sensing**, which had a special GIS issue in June of 1986. Two other important journals from Europe include **Geo-Processing**, which publishes articles on both concepts and procedures in GIS (Chrisman, 1984b; Clarke 1986; Eyton 1984; and Lorie and Meirer, 1984) and the **ITC Journal** (Enschede, Holland), which covers a wide variety of digital mapping topics including numerous articles on digital terrain models (Fredriksen et al. 1985; Makarovic 1984a; Makarovic 1984b; Oswald and Raelzsch 1984; Stefanovic and Sijmons 1984)

Computer journals also publish articles on GIS. Examples include **Computer Graphics and Image Processing** (Scheier 1981), and **Computer Graphics and Applications** (MacDonald and Crain 1985). The discipline orientated journals occasionally carry articles on the applications of GIS. These include the **Journal of Forestry** (Berry 1986), **Environmental Management** (Bailey 1988; Berry 1987a), **The Journal of Soil and Water Conservation** (Walsh 1985), **Bioscience** (Scott et al. 1987), **Area** (Green et al. 1985), and **The Science of the Total Environment** (Jensen and Christensen, 1986). Examples of other Journals include **Computers, Environment, and Urban Systems** (Clarke 1986), **Surveying and Mapping** (Dahlberg 1986; Dale 1986), **IEEE Transactions on Geoscience and Remote Sensing** (McKeown 1987), and **AI in Natural Resource Management** (Robinson et al. 1987).

Books

There is a shortage of books on GIS. The first textbook, **Principles of Geographical Information Systems for Land Resources Assessment** by Burrough (1986), has been well received. Related volumes that have been used as GIS textbooks include both **Computer-Assisted Cartography: Principles and Prospects** by Monmonier (1982) and **Computer Mapping: Progress in the '80s** by Carter (1984). Edited compendiums on GIS include **Computer Graphics and Environmental Planning** by Teicholz and Berry (1983); **Basic Readings in Geographic Information Systems** by Marble et al. (1984), and **Geographic Information Systems for Resource Management: A Compendium** by Ripple (1987).

Selected Bibliography

The following list of references presents recent publications on the fundamentals of Geographic Information Systems. This list is not intended to be all inclusive, but rather a selection of significant articles published from January 1982 through December, 1987. This bibliography is organized according to the following topics in GIS fundamentals: 1. GIS overviews, principles, and trends, 2. Data capture, 3. Data structure, 4. Data manipulation, 5. Digital terrain models, 6. Data quality, error assessment, and 7. Expert Systems.

1. GIS OVERVIEWS, PRINCIPLES, AND TRENDS

Blakemore, M. 1985. From Lineprinter Maps to Geographic Information Systems: a Retrospective on Digital Mapping. *Bulletin, Society of University Cartographers* 19(2):65-70.

Blakemore, M. 1986. Cartography and Geographic Information Systems. *Progress in Human Geography* 10(4):553-563.

Burrough, P.A. 1986. *Principles of Geographical Information Systems for Land Resource Assessment.* Oxford University Press, New York.

Carter, J.R. 1984. Computer Mapping: Progress in the '80s. *Resource Publications in Geography.* Association of American Geographers, Washington D.C.

Clarke, K.C. 1985. Geographic Information Systems: Definitions and Prospects. *Bulletin, Geographic and Map Division, Special Libraries Association* 142:12-18.

Clarke, K.C. 1986a. Recent Trends in Geographic Information Systems Research. *Geo-Processing* (3):1-15.

Clarke, K.C. 1986b. Advances in Geographic Information Systems. *Computers, Environment and Urban Systems.* 10(3/4):175-184.

Dale, P.F. 1986. Problems and Research Needs in the Presentation of Land Information. *Surveying and Mapping* 46(2):151-155.

Dangermond, J. 1984. Geographic Data Base Systems. *Technical Papers, American Society for Photogrammetry and Remote Sensing/American Congress on Surveying and Mapping, Fall Convention.* pp 201-211.

Estes, J.E. 1984. Improved Information Systems: A Critical Need. *Proceedings, 10th International Symposium on Machine Processing of Remotely Sensed Data.* Laboratory for Applications of Remote Sensing, Purdue University, pp 2-8.

Green, N.P., S. Finch and J.W. Wiggins. 1985. The 'State of the Art' in Geographical Information Systems. *Area* 17(4):295-301.

Jackson, M.J. and D.C. Mason. 1986. The Development of Integrated Geo-Information Systems. *International Journal of Remote Sensing* 7(6):723-740.

Loveland, T.R. and B. Ramey. 1986. *Applications of U.S. Geological Survey Digital Cartographic Products, 1979-1983,* USGS Bulletin 1583. Reston, VA: U.S. Geological Survey.

Marble, D.F., H.W. Calkins, and D.J. Peuquet. 1984. *Basic Readings in Geographic Information Systems.* SPAD Systems LTD, Williamsville, New York.

Monmonier, M.S. 1982. *Computer-Assisted Cartography, Principles and Prospects.* Prentice Hall, Inc., New Jersey.

Optiz, B.K. (ed.) 1986. *Geographic Information Systems in Government.* A. Deepak Publishing, Hampton, Virginia.

Parent, P. and R. Church. 1987. Evolution of Geographic Information Systems as Decision Making Tools. *Proceedings GIS 87. American Society for Photogrammetry and Remote Sensing.* San Francisco, CA. pp 63-70.

Peucker, T.K. 1980. Literature for Geographic Information Systems. *Urban, Regional, and State Government Applications of Computer Mapping.* Harvard Library of Computer Graphics, 11:175-179.

Peuquet, D. J. and O'Callaghan J. (eds.) 1983. *Proceedings of the United States/Australia Workshop on the Design and Implementation of Computer-Based Geographic Information Systems.* IGU Commission on Geographical Data Sensing and Processing. Amherst, New York.

Ripple, W.J. (ed.) 1987. *Geographic Information Systems for Resource Management: A Compendium.* American Society for Photogrammetry and Remote Sensing, Falls Church, VA., 288pp.

Robinove, C.J. 1986. *Principles of Logic and Use of Digital Geographic Information Systems*, USGS Circular 977. Reston, VA: U.S. Geological Survey.

Smith, T.R., S. Menon, J.L. Star and J.E. Estes. 1987. Requirements and Principles for the Implementation and Construction of Large-Scale Geographic Information Systems. *International Journal of Geographical Information Systems.* 1(1):13-31.

Star, J.L., M.J. Cosentino, and T.W. Foresman. 1984. Geographic Information Systems: Questions to Ask Before It's Too Late. *Proceedings of the 10th International Symposium on Machine Processing of Remotely Sensed Data.* Purdue University Laboratory for Applications of Remote Sensing, pp 194.

Stefanovic, P. and J. Drummond. 1987. Selection and Evaluation of Computer-Assisted Mapping and Geo-Information Systems. *ITC Journal.* 1987-1:39-44.

Teicholz, E. and B.J.L. Berry. 1983. *Computer Graphics and Environmental Planning.* Prentice-Hall, Englewood, Cliffs, New Jersey. 150pp.

Tomlinson, R.F. 1984. Geographic Information Systems - a New Frontier. *Proceedings of the International Symposium on Spatial Data Handling.* Zurich, Switzerland. pp 1-14.

Tomlinson, R.F. 1987. Current and Potential Uses of Geographical Information Systems: The North American Experience. *International Journal of Geographical Information Systems,* 1 (3), pp. 203-218.

White, M.S. Jr. 1984. Technical Requirements and Standards for a Multipurpose Geographic Data System. *The American Cartographer.* 11(1):15-26.

Wilkinson, G.G. and P.F. Fisher. 1987. Recent Development and Future Trends in Geo-Information Systems. *The Cartographic Journal.* 24:64-70.

2. DATA CAPTURE

Baraniak, D.W. 1986. Scanning Technology: A Review and Look Into the Future. *Proceedings: AM/FM International Eastern Regional Conference.* Williamsburg, Virginia. pp 47-53,

Boyle, R.A. 1983. The Status of Graphic Data Input to Spatial Data Handling Systems. *Proceedings, United States/Australia Workshop on Design and Implementation of Computer-Based Geographic Information Systems.* IGU Commission on Geographical Data Sensing and Processing. pp13-20.

Callahan, G.M. 1986. The Latest in Scanning Appliances at the USGS. *Proceedings: AM/FM International Eastern Regional Conference.* Williamsburg, Virginia. pp 79-83.

Dangermond, J. 1988. A Review of Digital Data Commonly Available and Some of the Practical Problems of Entering Them Into a GIS. *Technical Papers 1988 American Congress on Surveying and Mapping/American Society for Photogrammetry and Remote Sensing Annual Convention.* St. Louis, Missouri. 5:1-10.

Gwynn, Robert L. 1986. Collecting Geographic Information Systems Data On the Laser-scan Automated System at US Geological Survey. *Geographic Information Systems in Government.* Vol. II. pp 523-530.

Leberl, F.W. and D. Olson. 1982. Raster Scanning for Operational Digitizing of Graphical Data. *Photogrammetric Engineering and Remote Sensing.* 48(4):615-627.

Lorie, R.A. and A. Meirer. 1984. Using a Relational DBMS for Geographical Databases. *Geo-Processing.* 2:243-257.

Marble, D.F., J.P. Lauzon and M. McGranaghan. 1984. Development of a Conceptual Model of the Manual Digitizing Process. *Proceedings of the International Symposium on Spatial Data Handling.* Zurich, Switzerland. pp 146-171.

Opheim, H. 1982. Fast Data Reduction of a Digitized Curve. *Geo-Processing.* 2:33-40.

Rodrigues, M. and R.S. Baxter. 1982. An Approach to Geocoding in Less Developed Countries. *Geo-Processing.* 2(1):41-49.

Walker, A.S. 1987. Input of Photogrammetric Data to Geographical Information Systems. *Photogrammetric Record,* 12(70) pp. 459-471.

3. DATA STRUCTURE

Chrisman, N.R. 1984b. On Storage of Coordinates in Geographic Information Systems. *Geo-Processing* 2:259-270.

Cowen, D.J. 1983. Using Standard Output Formats for Distributed Geographical Data Handling. *Proceedings, United States/Australia Workshop on Design and Implementation of Computer-Based Geographic Information Systems*. IGU Commission on Geographic Data Sensing and Processing, pp133-138.

John, R.B. 1983. Data Structure Considerations in Topographic Mapping. *Proceedings, United States/Australia Workshop on Design and Implementation of Computer-Based Geographic Information Systems*. IGU Commission on Geographic Data Sensing and Processing, pp 21-28.

Mark, D.M. and J.P. Lauzon. 1984. Linear Quadtrees for Geographic Information Systems. *Proceedings of the International Symposium on Spatial Data Handling*. Zurich, Switzerland, pp. 412-430.

Palimaka, J., O. Halustchak and W. Walker. 1986. Integration of a Spatial and Relational Database within a Geographic Information System. *Proceedings of the American Society for Photogrammetry and Remote Sensing/American Congress on Surveying and Mapping Annual Convention*. Washington, DC. pp 131-140.

Peuquet, D.J. 1983. Vector/Raster Options for Digital Cartographic Data. *Proceedings United States/Australia Workshop on Design and Implementation of Computer-Based Geographic Information Systems*. IGU Commission on Geographic Data Sensing and Processing. pp 29-35.

Peuquet, D.J. 1984a. Data Structures for a Knowledge-Based Geographic Information System. *Proceedings of the International Symposium on Spatial Data Handling*. Zurich, Switzerland. pp 372-391.

Peuquet, D.J. 1984b. A Conceptual Framework and Comparison of Spatial Data Models. *Cartographica*. 21(4):66-113.

Peuquet, D.J. 1986. The Use of Spatial Relationships to Aid Spatial Database Retrieval. *Proceedings of the Second International Symposium on Spatial Data Handling*. Seattle, Washington. pp 459-471.

Samet, H. 1983. Hierarchical Data Structures for Representing Geographical Information. *Proceedings, United States/Australia Workshop on Design and Implementation of Computer-Based Geographic Information Systems*. IGU Commission on Geographic Data Sensing and Processing, pp 36-50.

Samet, H. 1984. The Quadtree and Related Hierarchical Data Structures. *ACM Computing Surveys*. 16(2):187-260.

Samet, H. and R.E. Webber. 1984. On Encoding Boundaries with Quadtrees. *IEEE Transactions on Pattern Analysis and Machine Intelligence*. PAMI-6(3):365-369.

Samet, H., A. Rosenfeld, C.A. Shaffer, and R.E. Weber. 1984. A Geographic Information System Based on Quadtrees. *Proceedings of the International Symposium on Spatial Data Handling*. Zurch, Switzerland, pp 393-411

Shneier, M. 1981. Two Hierarchical Linear Feature Representations: Edge Pyramids and Edge Quadtrees. *Computer Graphics and Image Processing*. 17(3):211-224.

Waugh, T.C. 1986. A Response to Recent Papers and Articles on the Use of Quadtrees for Geographic Information Systems. *Proceedings of the Second International Symposium on Spatial Data Handling*. Seattle, Washington. pp 33-37.

4. DATA MANIPULATION

Berry, J.K. 1986. Learning Computer Assisted Map Analysis. *Journal of Forestry* 84(10):39-42.

Berry, J.K. 1987a. A Mathematical Structure for Analyzing Maps. *Environmental Management* 11(3):317-325.

Berry, J.K. 1987b. Fundamental Operations in Computer-Assisted Map Analysis. *International Journal of Geographic Information Systems* 1(2):119-136.

Dangermond, J. 1983. A Classification of Software Components Commonly Used in Geographic Information Systems. *Proceedings, United States/Australia Workshop on Design and Implementation of Computer-Based Geographic Information Systems*. IGU Commission on Geographic Data Sensing and Processing, pp70-91.

Hopkins, L.d. 1977. Methods for Generating Land Suitability Maps: A Comparative Evaluation. *American Institute of Planners Journal*, pp. 386-400.

Jensen, J.R. and E.J. Christensen. 1986. Solid and Hazardous Waste Disposal Site Selection Using Digital Geographic Information System Techniques. *The Science of the Total Environment*, 56, pp. 265-276.

MacDonald, C. L. and I.K. Crain. 1985. Applied Computer Graphics in a Geographic Information System: Problems and Successes. *Computer Graphics and Applications.* 5(10):34-39.

Tomlin, C.D. 1983a. Digital Cartographic Modeling Techniques in Environmental Planning. Unpublished Ph.D. Dissertation, Yale University, Connecticut. 290pp.

5. DIGITAL TERRAIN MODELS

Fredriksen, P., O. Jacobi and K. Kubik. 1985. A Review of Current Trends in Terrain Modeling. *ITC Journal* 1985-2:101-106.

Makarovic, B. 1984a. Structures for Geo-Information and Their Application in Selective Sampling for Digital Terrain Models. *ITC Journal* 1984-4:285-295.

Makarovic, B. 1984b. Automated Production of DTM Data Using Digital Off-line Techniques. *ITC Journal.* 1984-2:135-141.

Mark, D.M. 1984. Automated Detection of Drainage Networks from Digital Elevation Models. *Cartographica.* 21(2/3):168-178.

Marks, D., J. Dozier and J. Frew. 1984. Automated Basin Delineation from Digital Elevation Data. *Geo-Processing.* 2:299-311.

Oswald, H. and H. Raetzsch. 1984. A System for Generation and Display of Digital Elevation Models. *Geo-Processing.* 2:197-218.

Stefanovic, P. and K. Sijmons. 1984. Computer Assisted Relief Representation. *ITC Journal.* 1984-1:40-47.

6. DATA QUALITY, ERROR ASSESSMENT

Bailey, R.G. 1988. Problems With Using Overlay Mapping for Planning and Their Implications for Geographic Information Systems. *Environmental Management* 12(1):11-17.

Blakemore, M. 1984. Generalization and Error in Spatial Data Bases. *Cartographica* 21(3/4):131-139.

Chrisman, N. R. 1984a. The Role of Quality Information in the Long Term Functioning of a Geographic Information System. *Cartographica* 21(2/3):79-87.

Cook, B.G. 1983. Geographic Overlay and Data Reliability. *Proceedings, United States/Australia Workshop on Design and Implementation of Computer-Based Geographic Information Systems*. IGU Commission on Geographic Data Sensing and Processing, pp 64-69.

Dahlberg, R.E. 1986. Combining Data from Different Sources. *Surveying and Mapping* 26(2):141-149.

Drummond, J. 1987. A Framework for Handling Error in Geographic Data Manipulation. *ITC Journal* 1:73-82.

Franklin, W.R. 1984. Cartographic Errors Symptomatic of Underlying Algebra Problems. *Proceedings of the International Symposium on Spatial Data Handling*. Zurich, Switzerland. pp 190-208.

Mead, D.A. 1982. Assessing Data Quality in Geographic Information Systems. In:*Remote Sensing for Resource Management*. pp 51-62.

Newcomer, J.A. and J. Szajgin. 1984. Accumulation of Thematic Map Errors in Digital Overlay Analysis. *The American Cartographer*. 11(1):58-62.

Wehde, M. 1982. Grid Cell Size in Relation to Errors in Maps and Inventories Produced by Computerized Map Processing. *Photogrammetric Engineering and Remote Sensing*. 48(8):1289-1298.

7. EXPERT SYSTEMS

Antony, R. and P.J. Emmerman. 1986. Spatial Reasoning and Knowledge Representation. In: B.K. Opitz (ed), *Geographic Information Systems in Government, Vol. 2*, A. Deepak Publishing, Hampton, Virginia, pp 795-814.

McKeown, D.M. 1987. The Role of Artificial Intelligence in the Integration of Remotely Sensed Data with Geographic Information Systems. *IEEE Transactions on Geoscience and Remote Sensing*. GE-25(3):330-348.

Ripple, W.J. and V.S. Ulshoefer. 1987. Expert Systems and Spatial Data Models for Efficient Geographic Data Handling. *Photogrammetric Engineering and Remote Sensing.* 53(10):1431-1433.

Robinson, G. and M. Jackson. 1985. Expert Systems in Map Design. *Proceedings of the 8th International Symposium on Computer-Assisted Cartography.* Baltimore, Maryland, pp. 430-439.

Robinson, V.B., A.U. Frank and H.A. Karimi. 1987. Expert Systems for Geographic Information Systems in Resource Management. *AI Applications in Natural Resource Management.* 1(1):47-57.

Smith, T.R. and M. Pazner. 1984. Knowledge-Based Control of Search and Learning in a Large-Scale GIS. *Proceedings of the International Symposium on Spatial Data Handling.* Zurich, Switzerland, pp 498-529.

Smith, T.R., D. Peuquet, and S. Menon. 1988. KBGIS-II: A Knowledge-Based Geographic Information System. *International Journal of Geographic Information Systems.* 1(2):149-172.

GEOGRAPHIC INFORMATION SYSTEMS NEWSLETTERS

by

James W. Merchant and Loyola M. Caron*

U.S. - NATIONAL COVERAGE

AM/FM SCRIBE**
AM/FM International
Administrative Office
c/o Barbara Emery
8775 East Orchard Road
Suite 820
Englewood, CO 80111
Editor: Ginger M. Juhl
Telephone: 303-779-8320

CARTOGRAPHIC PERSPECTIVES**
North American Cartographic Information Society
6010 Executive Boulevard, Suite 100
Rockville, MD 20852
Telephone: 301-443-8075
Editors: David DiBiase and Karl Proehl
Pennsylvania State University
Department of Geography
302 Walker Building
University Park, PA 16802
Telephone: 814-865-3433

---

* The authors may be contacted at the Center for Advanced Land Management Information Technologies (CALMIT), Conservation and Survey Division, University of Nebraska-Lincoln, 113 Nebraska Hall, Lincoln, Nebraska 68588-0517; telephone: 402-472-2567. Information regarding additions or corrections would be welcome.

** Membership in national society is/may be required.

NOTE: A list of newsletters dealing with remote sensing is also available. A copy may be obtained by contacting the authors.

DIGITAL MAPPING, CHARTING AND GEODESY ANALYSIS
  PROGRAM (DMAP) NEWS
      Naval Ocean Research and Development Activity
        (NORDA)
      Code 351
      John C. Stennis Space Center, MS 39529-5004
      Editor: Susan Carter
      Telephone: 601-688-4652

ESIC NEWSLETTER
      Earth Science Information Center
      Department of the Interior
      U.S. Geological Survey
      507 National Center
      Reston, VA 22092
      Editor: Pat L. Creech
      Telephone: 703-860-6045

F-M (Facilities Management) AUTOMATION NEWSLETTER
      18030 Polvera Way
      San Diego, CA 92128-1123
      Editor: Chris Harlow
      Telephone: 619-451-2995
      Subscription: $125/year - U.S. (12 issues)
                     $175/year - overseas

FDC NEWSLETTER
      Federal Interagency Coordinating Committee
        on Digital Cartography
      Department of the Interior
      U.S. Geological Survey
      516 National Center
      Reston, VA 22092
      Editor: Bruce McKenzie
      Telephone: 703-648-5740

GIMS (GEOGRAPHIC INFORMATION MANAGEMENT SYSTEMS)
  NEWS
      GIMS Committee
      American Society for Photogrammetry and
        Remote Sensing/American Congress on
        Surveying and Mapping
      210 Little Falls Street
      Falls Church, VA 22046
      Telephone: 703-534-6617
      Editor: Paul Durgin
               1 Old Dover Rd., Suite 5
               Rochester, NH 03867
               Telephone: 603-332-6833

GIS/CADD MAPPING SOLUTIONS
    Venture Communications, Inc.
    P.O. Box 02332, Department K
    Portland, OR 97202
    Editor: Michael J. Carey
    Telephone: 503-236-5810
    Subscription: $99/year - U.S. (12 issues)
                    $124/year - all others

THE GIS FORUM
    THG Publishing Company
    16306 Sir William Drive
    Spring, TX 77379-7638
    Editor: Francis L. Hanigan
    Telephone: 800-GIS-4ALL
    Subscription: $125/year - U.S.;
                   $175/year - international

GIS NEWS
    Association of American Geographers
    GIS Specialty Group
    c/o University of South Carolina
    Humanities and Social Sciences Computer Lab
    Columbia, SC 29208
    Editor: David Cowen
    Telephone: 803-777-7840

GIS REVIEW
    Land Systems Corporation
    P.O. Box 496
    Greenland, NH 03840
    Editors: Gene Roe and Karen Keicher
    Telephone: 603-436-9538
    Subscription: $4.00/year (4 issues)

GIS WORLD
    GIS World, Inc.
    P.O. Box 8090
    Ft. Collins, CO 80526
    Editor: H. Dennison Parker
    Telephone: 303-484-1973
    Subscription: $96/year - U.S.;
                   $108/year - Canada;
                   $124/year - all others

THE ILIAD
        Institute for Land Information
        8th Floor
        440 1st Street NW
        Washington, DC 20001
        Editor: Doug Wilcox
                 Department of Civil Engineering
                 Box 3CE
                 New Mexico State University
                 Las Cruces, NM 88003
                 Telephone: 505-646-3801

MSFWIS NEWSLETTER
        Multi-State Fish and Wildlife Information
           Systems
        108 Colony Park
        2001 South Main Street
        Blacksburg, VA 24060
        Editors: Jeff Waldon and Karen Kelly
        Telephone: 703-231-7348

NCGIA UPDATE
        National Center for Geographic Information
           and Analysis
        University of California-Santa Barbara
        Santa Barbara, CA 93106
        Editor: Phil Parent
        Telephone: 805-961-4617 or 8224

REMOTE SENSING AND GEOGRAPHIC INFORMATION SYSTEMS
   STATUS REPORT
        U.S. Department of Agriculture
        Soil Conservation Service
        P.O. Box 6567
        Fort Worth, TX 76115
        Contact: Wayne Weaver
        Telephone: 817-334-5212

SORSA NEWS**
        Spatially-Oriented Referencing Systems
           Association
        P.O. Box 3825, Station "C"
        Ottawa, Ontario K1Y 4M5
        CANADA
        Editor: Larry Li
        Telephone: 613-951-6921

URISA NEWS**
        Urban and Regional Information Systems
          Association (URISA)
        900 Second Street, NE
        Suite 304
        Washington, DC 20002
        Editor: Rebecca Somers
        Telephone: 202-289-1685

U.S. - STATE AND REGIONAL COVERAGE

    ARIZONA LAND RESOURCE INFORMATION SYSTEM (ALRIS)
      QUARTERLY NEWSLETTER
        Arizona Land Resource Information System
        State Land Department
        1624 West Adams
        Phoenix, AZ 85007
        Contact: Gary Irish
        Telephone: 602-542-4061

    CALMIT NEWS
        Center for Advanced Land Management
          Information Technologies (CALMIT)
        Conservation and Survey Division
        113 Nebraska Hall
        University of Nebraska-Lincoln
        Lincoln, NE 68588-0517
        Editor: James Merchant
        Telephone: 402-472-7531 or 2567

    GIS NEWS LAYERS
        Geographic Information Systems Unit
        Valuation Services Bureau
        Division of Equalization and Assessment
        Executive Department
        Sheridan Hollow Plaza
        16 Sheridan Avenue
        Albany, NY 12210-2714
        Editor: Marty Goldblatt
        Telephone: 518-474-3453

    GIS REPORT
        New York State Office of Parks, Recreation
          and Historic Preservation
        Bureau of Planning and Research
        Empire State Plaza
        Agency Building #1
        Albany, NY 12238
        Editor: Kathy B. Cramer
        Telephone: 518-474-0414

ILLINOIS GIS UPDATE
Illinois Geographic Information System
272 Natural Resources Building
607 East Peabody Drive
Champaign, IL 61820
Contact: Warren Brigham
Telephone: 217-333-8907

ILLINOIS MAP NOTES
Illinois State Geological Survey
Natural Resources Building
615 East Peabody Drive
Champaign, IL 61820
Editor: Richard E. Dahlberg
         Laboratory for Cartography and
            Spatial Analysis
         Department of Geography
         Northern Illinois University
         DeKalb, IL 60115
         Telephone: 815-753-0631

KANSAS APPLIED REMOTE SENSING (KARS) NEWSLETTER
Kansas Applied Remote Sensing Program
University of Kansas Space Technology Center
2291 Irving Hill Drive
Lawrence, KS 66045-2969
Editor: Edward Martinko
Telephone: 913-864-7720

LMIC NEWS
Land Management Information Center (LMIC)
Minnesota State Planning Agency
300 Centennial Building
658 Cedar Street
St. Paul, MN 55101
Editor: Laura Muessig
Telephone: 612-296-1208

MASS-GIS NEWSLETTER
MASS-GIS Project
80 Boylston Street, Suite 955
Boston, MA 02116
Editor: David Weaver
Telephone: 617-727-6930

NEW JERSEY GIS UPDATE
        Geographic and Statistical Analysis Unit
        Division of Science and Research
        Department of Environmental Protection
        401 East State Street, 6th Floor
        Trenton, NJ 08625
        Editor: Barbara Plunkett
        Telephone: 609-633-2641

NRGIS NEWS
        Minnesota Natural Resource Geographic
          Information
        System Consortium (NRGIS)
        Land Management Information Center
        Minnesota State Planning Agency
        300 Centennial Office Building
        658 Cedar Street
        St. Paul, MN 55155
        Editor: Charlie Parson
        Telephone: 612-296-1211

OREGON CURSOR
        Oregon Water Resources Department
        3850 Portland Rd. NE
        Salem, OR 97310
        Editor: Rick Bastasch
        Telephone: 503-378-3671

RHODE ISLAND GIS (RIGIS) NEWS
        Department of Natural Resources Science
        Environmental Data Center
        Woodward Hall
        The University of Rhode Island
        Kingston, RI 02881-0804
        Editor: Peter August
        Telephone: 401-792-4794

TEXAS NATURAL RESOURCES INFORMATION SYSTEM
   NEWSLETTER
        Texas Natural Resources Information System
        P.O. Box 13231
        Austin, TX 78711-3231
        Contact: Laverne Willis
        Telephone: 512-463-8337

WISCONSIN LAND INFORMATION NEWSLETTER
    Center for Land Information Studies
    1040 WARF Building, 610 Walnut Street
    Institute for Environmental Studies
    University of Wisconsin-Madison
    Madison, WI   53705
    Editor: Bernard J. Niemann, Jr.
    Telephone: 608-263-6843

WISCONSIN MAPPING BULLETIN
    State Cartographer's Office
    155 Science Hall
    University of Wisconsin-Madison
    Madison, WI   53706-1404
    Editors: Art Ziegler and Bob Gurda
    Telephone: 608-262-3065

CANADA

CANADIAN CARTOGRAPHIC ASSOCIATION NEWSLETTER**
    Canadian Cartographic Association
    c/o Department of Geography
    Memorial University of Newfoundland
    St. John's, Newfoundland   A1B 3X9
    CANADA
    Telephone: 709-737-8988

FEDERAL GEOMATICS BULLETIN
    Secretariat, IACG
    Geographic Information Systems Division
    Energy, Mines and Resources Canada
    615 Booth Street
    Ottawa, Ontario   K1A 0E9
    CANADA

LOCUS FOCUS
    Ontario Ministry of Natural Resources
    Geographic Information Services
    4th Floor
    90 Sheppard Avenue East
    North York, Ontario   M2N 3A1
    CANADA

<u>LRIS (LAND-RELATED INFORMATION SYSTEMS)</u>
    <u>NEWSLETTER</u>
        Land-Related Information Services Group
        Bureau of Surveying and Mapping
        Alberta Forestry, Lands and Wildlife
        910 108th Street Building
        9942 -108th Street, NW
        Edmonton, Alberta T5K 2J5
        CANADA
        Editor: L.T. Toth
        Telephone: 403-422-1413

AFRICA, ASIA, AUSTRALIA and EUROPE

    <u>ALICNEWS</u>
        Australian Land Information Council
        P.O. Box 2
        Belconnen ACT 2616
        AUSTRALIA
        Telephone: (062) 526984

    <u>AURISA NEWS</u>**
        Australian Urban and Regional Information
            Systems Association
        G.P.O. Box 312
        Eastwood 2122, N.S.W.
        AUSTRALIA
        Editor: Ken Bullock
        Telephone: SYDNEY (02) 265-9506
        Subscription: $20/year, quarterly

    <u>AUSTRALIAN KEY CENTRE IN LAND INFORMATION STUDIES</u>
        <u>(AKCLIS) NEWSLETTER</u>
        Queensland University of Technology
        GPO Box 2434
        Room 305 'H' Block
        Brisbane, Queensland 4001
        AUSTRALIA

BURISA NEWSLETTER**
        British Urban and Regional Information
          Systems Association
        c/o Susan Leather
        School for Advanced Urban Studies
        University of Bristol
        Rodney Lodge
        Grange Road
        Bristol, England BS8 4EA
        Telephone: (0272) 741117
        Subscription: 9.75 pounds - U.K.
                       (5 issues/year)
                       10.50 pounds - overseas

CONTAGIS
        Department of Surveying
        University of the Witwatersrand
        PO WITS
        Johannesburg 2050
        REPUBLIC OF SOUTH AFRICA
        Editor: Russell M. Davies

GRID NEWS
        Global Resource Information Database
        Global Environmental Monitoring System
        P.O. Box 30552
        Nairobi, Kenya

LANDFORM
        Land Information Branch
        Department of Lands
        144 King William Street
        GPO Box 1047
        Adelaide, South Australia 5001
        AUSTRALIA

LINC NEWSLETTER
        Commonwealth Land Information Support Group
          (CLISG)
        P.O. Box 2
        Belconnen ACT 2616
        AUSTRALIA
        Contact: The Director
        Telephone: (062) 526984

MAPPING AWARENESS
Miles Arnold
High Winds
Cassington
Oxford, England OX8 1DL
Editor: Peter J. Shand
Telephone: OXFORD (0865) 880236
Subscription: (Bi-monthly) UK - 55 pounds;
                Overseas Airmail - 65 pounds

SLIC NEWSLETTER
State Land Information Council Directorate
Department of Lands Building
G.P.O. Box 39
23-33 Bridge Street
Sydney, New South Wales 2001
AUSTRALIA
Telephone: (02) 228 6111

USEMAP NEWSLETTER
Department of Urban Survey and Human
  Settlement Analysis
International Institute for Aerial Survey
  and Earth Sciences (ITC)
P.O. Box 6, 7500 AA
Enschede, The Netherlands
Editor: C.A. De Bruijn

NEWSLETTERS FOCUSED ON VENDORS OF SPECIFIC SOFTWARE OR GIS SERVICES

ARC NEWS
Environmental Systems Research Institute
380 New York Street
Redlands, CA 92373
Telephone: 714-793-2853

AUTOGIS NEWSLETTER
Autometric Inc.
165 South Union Boulevard
Suite 902
Lakewood, CO 80228-2214
Contact: Bruce Morse
Telephone: 303-989-6377

GDT NEWS
Geographic Data Technology Inc.
13 Dartmouth College Highway
Lyme, NH 03768
Telephone: 603-795-2183

GEOBASED SYSTEMS USERS GROUP NEWSLETTER

GeoBased Systems
4800 Six Forks Road
Raleigh, NC 27609
Contact: Carol Hammond
Telephone: 919-783-8000

GEOFORUM

Strategic Locations Planning
4030 Moorpark Avenue
Suite 123
San Jose, CA 95117-1848
Editor: Thomas Cook
Telephone: 408-985-7400

GIMMS NEWSLETTER

GIMMS Ltd
30 Keir Street
Edinburgh
Scotland, U.K. EH3 9EU
Editor: Marlene A. Ferenth
Telephone: 031-668-3046

GRASSCLIPPINGS

U.S. Army Construction Engineering Research
   Laboratory (USA-CERL)
ATTN: ENDIV, Mary Martin
P.O. Box 4005
Champaign, IL 61820-1305
Editors: Mary Martin and Kathryn Norman
Telephone: 1-800-252-7122 x 220
               (Inside Illinois)
            1-800-USA-CERL x 220
               (Outside Illinois)
Publisher: GRASS Interagency Steering
               Committee
            Institute for Technology
            Development
            ATTN: Missy Escher
            Space Remote Sensing Center
            Building 1103, Suite 118
            John C. Stennis Space Center,
               MS 39529

HUNTER GIS FORUM
    Hunter GIS, Inc.
    Suite 170
    1121 Wood Ridge Center Drive
    Charlotte, NC 28217
    Contact: Pamela McCray
    Telephone: 704-525-8620

IIS REVIEW
    Integrated Information Systems
    450 B Street, Suite 1200
    San Diego, CA 92112
    Contact: Jeff Sauter
    Telephone: 619-232-7008

INTERACTIONS
    Synercom Technology, Inc.
    2500 City West Boulevard
    Suite 1100
    Houston, TX 77042
    Editor: Vickie Williams
    Telephone: 713-954-7000

INTERVUE
    Intergraph Corporation
    One Madison Industrial Park
    Marketing Communications
    Mail Stop IW1510
    Huntsville, AL 35807-4201
    Editor: Jeannie Robison
    Telephone: 205-772-2700

THE MONITOR
    ERDAS, Inc.
    2801 Buford Highway
    Suite 300
    Atlanta, GA 30329
    Editor: James D. Mathis
    Telephone: 404-248-9000

PERSPECTIVE
    PAMAP Graphics Ltd.
    Suite 301-3440 Douglas Street
    Victoria, British Columbia V8Z 3L5
    CANADA
    Contact: Peter Archibald
    Telephone: 604-381-3838

QC DATA COLLECTORS NEWS
QC Data Collectors
777 Grant
Denver, CO 80203
Contact: Sara Roberts
Telephone: 303-837-1444

REMOTE SENSING AND DATABASE DEVELOPMENT NEWS
James W. Sewall Company
P.O. Box 433
147 Center Street
Old Town, ME 04468
Editors: Mark A. Jadkowski and Marian
          S. Dressler
Telephone: 207-827-4456
FAX: 207-827-3641

S9 ! PIPELINE
Prime Wild GIS, Inc.
373 Inverness Drive South
Englewood, CO 80112
Contact: Mike White

TYDAC NEWS
TYDAC Technologies, Inc.
Suite 310
1600 Carling Avenue
Ottawa, Ontario K1Z 8R7
CANADA
Telephone: 613-722-7472

ULTIMAP UPDATE
UltiMap Corporation
2901 Metro Drive, Suite 314
Minneapolis, MN 55425
Editor: Jerry Robinson
Telephone: 612-854-2382